Teacher Guide

Graphs and Rates

Murray Britt
Peter Hughes
Randall Souviney

Dale Seymour Publications®
Parsippany, New Jersey

Project Editor: Dottie McDermott
Managing Editor: Carolyn Coyle
Production/Manufacturing Director: Janet Yearian
Production/Manufacturing Manager: Karen Edmonds
Art Director: Jim O'Shea
Text and Cover Design: Robert Dobaczewski

Dale Seymour Publications
An imprint of Pearson Learning
299 Jefferson Road, P.O. Box 480
Parsippany, New Jersey 07054-0480
www.pearsonlearning.com
1-800-321-3106

ISBN 0-7690-2526-9

2 3 4 5 6 7 8 9 10-MZ-04 03 02 01 00

Contents

Program Overview

PreAlgebra Makes Sense is designed to provide middle-grade students with an understanding of the important prerequisite mathematical ideas needed for later success in algebra. As the first of a two-part series, *PreAlgebra Makes Sense* lays the foundation for algebraic thinking. *Algebra Makes Sense,* the second part of the series, will use this foundation to provide meaningful experiences with key algebraic ideas.

Consisting of six student books and six corresponding teacher guides, *PreAlgebra Makes Sense* presents the topics listed below in an interactive format that asks questions to make students think, stresses the process that leads to the answer, and encourages students to discuss the reasoning behind their answers.

BOOK 1: Fractions

Section 1: Investigating Fractions of Sets
Section 2: Investigating Fractions of Fractions
Section 3: Investigating Addition and Subtraction of Fractions
Section 4: Investigating Multiplication and Division of Fractions

BOOK 2: Signed Numbers and Powers

Section 1: Investigating Addition of Negative Numbers
Section 2: Investigating Subtraction of Negative Numbers
Section 3: Investigating Multiplication and Division of Negative Numbers
Section 4: Investigating Order of Operations
Section 5: Investigating Number Laws
Section 6: Investigating Powers

BOOK 3: Patterns of Factors and Multiples

Section 1: Investigating Factors and Multiples
Section 2: Investigating Factor Rules
Section 3: Investigating Common Multiples and Factors

BOOK 4: Number Investigations

Section 1: Investigating Fractions
Section 2: Investigating Calendar Numbers
Section 3: Investigating Consecutive Sums
Section 4: Investigating Arithmagons
Section 5: Investigating Number Boxes

BOOK 5: Graphs and Rates

Section 1: Investigating and Interpreting Graphs
Section 2: Investigating Measurement Conversion Rates
Section 3: Investigating Exchange Rates
Section 4: Investigating Percentage Rates
Section 5: Investigating Graphs and Changing Rates

BOOK 6: Measurement Patterns and Formulas

Section 1: Investigating Area Formulas
Section 2: Investigating Volume Formulas
Section 3: Investigating Circles and Cylinders

Developing Algebraic Thinking

PreAlgebra Makes Sense introduces students to the underlying ideas of algebraic thinking. Students draw diagrams and graphs and use their own language, as well as the language of mathematics, to express quantitative relationships. They use the results of these explorations to help solve a variety of intriguing mathematical problems. Students learn to work with numbers in flexible ways that help develop number sense. They invent their own shortcut methods for making sensible judgments about problems that are based on numerical relationships. This process of devising and generalizing these shortcut rules and formulas is central to students' development of algebraic thinking.

Through this process of generalizing, it is hoped that students will develop a deeper awareness of the power of mathematics. For example, in the formula for finding the area of a triangle, $Area = \frac{1}{2} \times base \times height$, the use of the indefinite article "a" indicates that the formula is true for *all* triangles, not just the ones that may have been selected as convenient examples. An important goal of the activities in *PreAlgebra Makes Sense* is to help students understand why such a seemingly simple statement about the relationship between parts of a triangle has such awesome mathematical implications.

Students must learn to generalize if they are to develop a true understanding of mathematical ideas. Students will need to investigate new ideas by themselves, in small groups, and in whole-class settings. They will need support from their teacher to draw out and consolidate the big mathematical ideas that emerge. Students will also need time to consolidate their understanding through ongoing exploration, clarifying discussion with other students and adults, and appropriate skill practice. It is for these reasons that the series has been subtitled *Interactive Tasks for Algebra Learners.*

Using PreAlgebra Makes Sense

PreAlgebra Makes Sense may be used in a variety of ways to help learners more fully understand the key mathematical concepts needed to study algebra successfully. Throughout the series, students are encouraged to work in small groups as they explore patterns and develop techniques to solve problems. The books may be used to supplement regular instruction or as replacement units for traditional textbook units when appropriate.

Using the Student Books—An Interactive Approach

Each section of the student book includes a series of tasks. Students are initially presented with a situation in a Task Box, which identifies the goal of the set of exercises that follow. For example, the Task Box below lets students know that they are going to be following a formula to convert Japanese yen to US dollars.

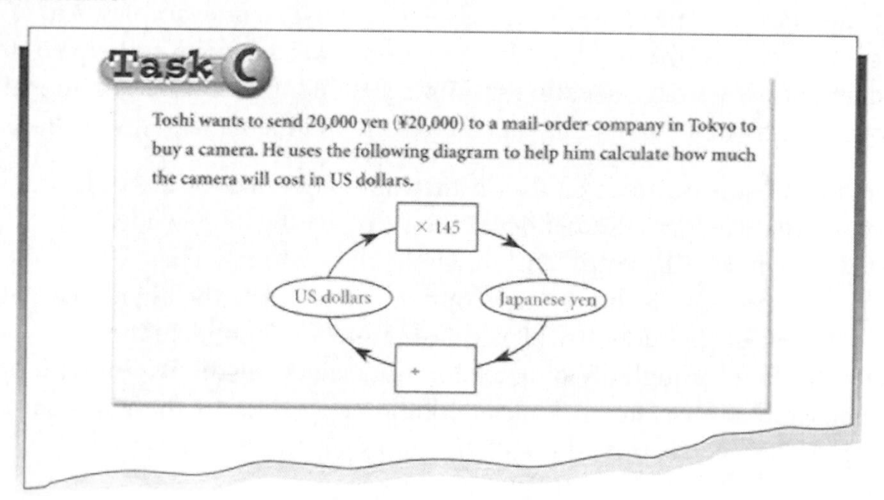
Task C

Toshi wants to send 20,000 yen (¥20,000) to a mail-order company in Tokyo to buy a camera. He uses the following diagram to help him calculate how much the camera will cost in US dollars.

Once students know the task, they turn their focus to a carefully constructed set of questions that guides them to complete the task.

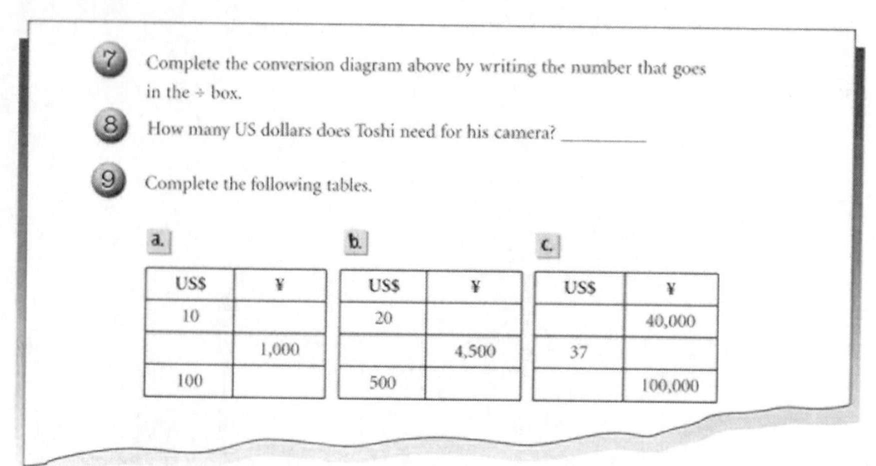
7. Complete the conversion diagram above by writing the number that goes in the ÷ box.
8. How many US dollars does Toshi need for his camera? ________
9. Complete the following tables.

a.

US$	¥
10	
	1,000
100	

b.

US$	¥
20	
	4,500
500	

c.

US$	¥
	40,000
37	
	100,000

For additional practice or for assessment of key concepts and ideas, Practice Plus pages are provided at the end of the book. Icons at particular exercises will reference these pages. Cumulative Practice is also included to help students consolidate prior skills.

Using the Teacher Guide

Each of the six student books in the *PreAlgebra Makes Sense* series has an accompanying teacher guide that includes mathematics goals as well as overviews of key mathematics topics and instructional approaches. Also included are teaching suggestions and solutions for all exercises and problems.

Throughout the teacher guide, teachers are encouraged to organize instruction so students are able to talk about their work with one another and explain ideas in their own words prior to teacher-led summary discussions. These prealgebra investigations have the greatest impact on student learning when teacher-led class discussion comes after students have had direct, independent experience with the mathematical ideas presented in each task.

Teaching and Assessing

The diagram below shows an instruction cycle for using *PreAlgebra Makes Sense* in the classroom. In this example, students begin by independently completing an exercise using a visual model. This experience is followed by discussion of the procedure with a partner or small group. Individuals then return to the book to extend the procedure they have created by completing assigned exercises. Finally, the teacher may lead a whole-class discussion to help students correct misunderstandings, extend procedures for all possible cases, and consolidate learning, with further practice in class or at home. This approach also provides valuable opportunities for teachers and parents to assess student progress.

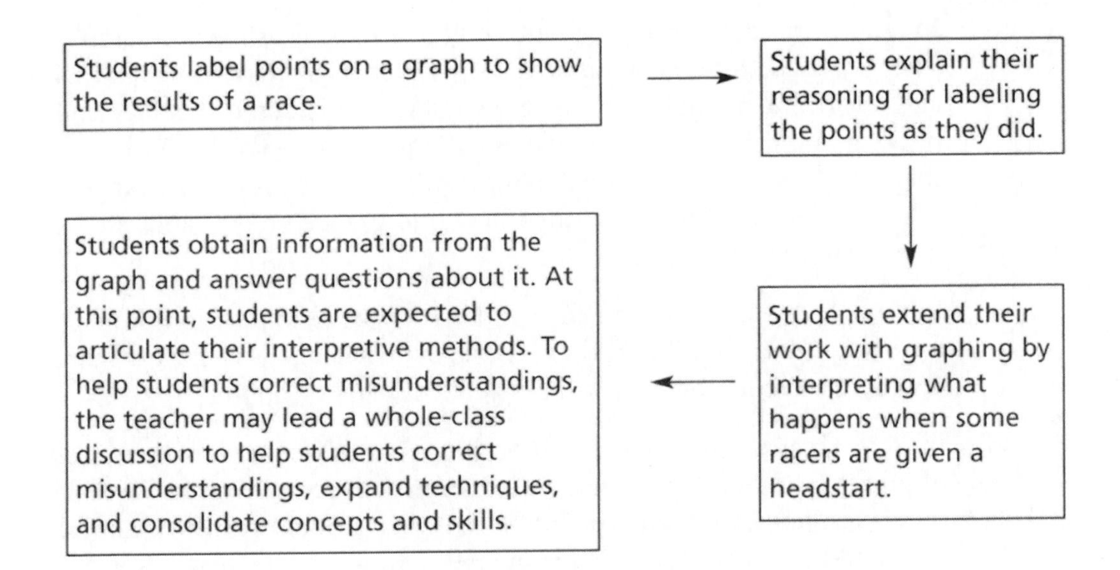

Investigating and Interpreting Graphs pages 5–19

GOALS

To compare information shown on a graph

To use graphs to help determine rates

Overview

This section introduces students to the idea of graphs without numbers. They learn to interpret information given on the graph by focusing on the graph itself rather than on scales of numbers that can be attached to a graph. As students compare two or more pieces of data, they learn to focus on the relationships between the data and to determine the location of an additional piece of data. Eventually, students work with numerical graphs and apply the graphing skills they have been working with to numerical situations.

Task A pages 5–7

In Item 1, students are asked to interpret relationships shown on graphs. They are asked to use comparative terms such as *lighter*, *heavier*, *shorter*, and *taller*. The intent here is to help students understand the direction of the graphical scales and how they interact with each other rather than graphing specific ordered pairs as points. The notion of change inherent in this approach is fundamental to the idea of variable, an important issue in algebra.

In Item 2, students use the Guess and Test strategy to position points for Alvin, Kim, and Chuck. In this case, the challenge is to have all three points coordinate correctly rather than just pairs of points.

In Item 3, students are asked to describe the relationship between pairs of points. This is extended to 3 points in Exercise 3c and most likely will pose a challenge to many students. Encourage students to work carefully through these tasks to help them grasp the fundamental features of graphs. In Item 4, students determine the better buy between two products.

Task B pages 7–8

In this task, students determine the best price among three products by using the information shown on graphs.

In Item 5, students first interpret a graph to determine the labels for the large-, the medium-, and the small-size can. The graph shows that one size B can costs slightly less than two size A cans and also weighs more (that is, its contents are greater) than two size A cans. In terms of weight, three size C cans equal two size B cans, which equal five size A cans. Also, three size C cans cost less than two size B cans, which in turn cost less than five size A cans. Students will benefit from making analyses and decisions about best-price deals. They will begin to develop an intuitive understanding of rate and also rate of change.

Task C pages 9–10

Here, the intuitive responses to the exercises in Items 9–13 are made more explicit by drawing a vertical line on the graph. This helps to provide a quick, visual way to compare prices for a particular weight. Point out to students that this is the concept behind unit prices that supermarkets use to help customers make decisions about which price offers the best deal.

As students work through Item 11, they need to locate a sensible position on the graph for an additional company, Sycamore Foods. This activity provides an opportunity to assess student understanding of the graphing concepts presented so far in this section.

Rather than a vertical line, a horizontal line can also indicate the best price by showing the greatest volume for a given price. In the graph for Item 12, the horizontal line indicates that for the same money, the greatest quantity is provided by product C and that product A is a better value than product B.

In Item 13, students are asked to demonstrate their understanding of rate by locating point D anywhere in the region formed by the angle AOC where O is the origin (unmarked) of the graph. A line from the origin through D intersects the gray horizontal line between the points where it intersects the lines from the origin through A and C.

Task D pages 11–14

Item 14 introduces numerical values on the graph axes for the first time. Students are first asked to identify each point, using the horizontal line at the 200-meter point on the axis. Then, in Item 15, students interpret the graph, using the 10-second vertical line. This activity provides an intuitive introduction to the idea of slope as it pertains to graphs. Encourage students to try and state the concept in their own words before you introduce the actual term. Then you may want to explain that the steepness of each line shows its slope.

In Item 17, students are challenged to extend the situation in Item 16 by adjusting head starts for Janine and Annie so that all race participants will finish at the same point. While their speeds, or rates, will not change (the new lines for J and A are parallel to the original lines), J's head start should be decreased and A's head start increased so that the points for J, C, and A on the 50-meter horizontal line will coincide. This will challenge most students and will require them to apply what they have learned so far in this section. Remind students to think about the slope of the lines.

The triathlon race represented by the graph in Item 18 also requires students to further conceptualize the idea of slope as an indicator of rate (speed in this case). This activity culminates in students' assigning value scales to the time and distance axes.

Task E pages 15–18

This task provides opportunities to consolidate the concepts related to rate illustrated in graphic models. In Items 20 and 21, students write short sentences to interpret distance-time graphs. This is a skill that can be difficult for many students. You can use Items 22 and 23 to assess students' understanding of this topic.

Task F pages 18–19

In the final task of this section, students are asked not only to conceptually interpret the graphic situations shown in the graph but also to determine specific rates (speeds). For example, Item 24 states that Larry is traveling at 50 miles per hour while the graph shows that his friend travels at a speed (rate) of 25 miles in 45 minutes, which is about 33 miles per hour. In Item 25, students are asked to extend the graphs in order to predict home arrival times for both drivers.

Section 2

Investigating Measurement Conversion Rates

pages 20–26

GOALS

To calculate rates, using graphs and patterns

To use conversion diagrams to convert between and among customary and metric units of length and weight

Overview

Students use the observations they make while studying a graphic representation of data to calculate rates and to form generalizations. They use diagrams to convert units of length and weight—kilometers to miles, miles to kilometers, pounds to kilograms, and kilograms to pounds.

Task A

pages 20–21

In this task, students find particular values represented on graphs and use patterns to help calculate rates. Such graphs and the conversion diagrams (flowcharts), like the one in Item 2, provide valuable visual cues to help students convert distance measurements from miles to kilometers and from kilometers to miles.

Task B

page 22

In Task B the earlier conversion diagram is extended to convert miles to centimeters, and then to meters and kilometers. The table in Item 7 provides an opportunity for students to consolidate their understanding by using the conversion diagram from left to right as well as from right to left.

Task C

pages 23–24

In Task C, students use a similar approach to convert from kilograms to pounds and from pounds to kilograms.

Task D

pages 25–26

Here the rate conversions lead to the development of a simple rule for determining appropriate distances when driving behind another car.

The 2-second following rule means, for example, that if you are traveling at 44 mph, the distance between your car and the car in front of you should be about 44 yards. In Item 16, students are asked to express this generalization in their own words. Writing generalizations such as this is central to algebraic thinking.

Section 3

Investigating Exchange Rates

pages 27–32

GOALS

To convert between two different currencies, using a graphic representation

To recognize the relationships among the values of currencies and their rates of exchange

Overview

Students continue to work with rates using graphs and then use conversion diagrams to convert among different currencies. Throughout this section, students gain experience focusing on the relationships among the values of these currencies.

Task A

pages 27–28

Exchange rates also offer another opportunity to work with rates. In Task A, students read graph coordinates to convert New Zealand dollars (NZ) to United States dollars (US).

Task B

pages 28–29

Here, students calculate the exchange rate between the two currencies used in Task A by using a conversion diagram similar to those developed in Section 2. The currency conversion diagram (flowchart) provides a visual way to solve the following.

NZ dollars $\times$ 0.55 = US dollars

So, US dollars $\div$ 0.55 = NZ dollars

Help students see that the diagram shows the inverse relationship between multiplication and division. However, the exchange rate is the number used with multiplication to convert one currency to another.

Therefore, the exchange rate for converting US dollars to NZ dollars is 1.82, not 0.55.

Task C pages 29–30

Here, students complete the conversion diagram without the support provided by a graph. Opposite operations are used (× and ÷) to convert dollars to yen and yen to dollars, so the values in the boxes are identical.

Task D page 30

In this task, students must apply the ideas developed previously. Tasks such as these and the subsequent tasks E and F further consolidate students' understanding of rate as applied to exchange rates.

Tasks E and F pages 31–32

In these two tasks, students must link three currency exchange rates, french franc (FF), British pound (£), and United States dollar (US$) by means of the currency conversion diagrams. For example, Item 19 challenges students to explain the reasoning behind the relationship FF = British pounds × 1.72 × 6.39. (Here, the intermediary relationships are US$ = £ × 1.72 and FF = US$ × 6.39.)

Section 4 Investigating Percentage Rates pages 33–37

GOALS

To determine the sale price at a given discount

To determine the original price given the sale price and rate of discount

Overview

Students discover ways of calculating various rates, such as discount, population increase, and sales tax. They move from using an intuitive, mental-math approach (as in Task A) to interpreting diagrams that help them generalize how multiplication and division are used for calculating rates.

Task A

page 33

In this task, Bela develops a clever way to calculate the amount of discount. She recognizes that 1% of $85 is $0.85 (1% of $1 is 1¢) so, 12% of $85 can be calculated as 12 × 85¢, or 10% of $85 + $0.85 + $0.85 = $8.50 + $0.85 + $0.85 = $10.20. Students are often motivated by challenges to find other clever ways to calculate such discounts. This kind of flexible thinking, where students invent and generalize for themselves, is fundamental to algebraic thinking.

Task B

pages 34–35

In this task, students should begin to realize that a sale discount of 12% means that the sale price is 88% of the original price. 88% = 0.88, so the sale price = 0.88 × original price.

In Item 6, students have an opportunity to consolidate the notion of discount so that they see that a discount of 15% off is the same as paying 85% of the original price.

Task C

pages 35–36

A 5% tax means that Sandy has to pay the original price plus 5% of the retail price. So she pays 100% (all) of the original price plus another 5% of the original price, which is 105% of the original price. This approach combines the two steps involved in first finding the amount of tax, and then adding it to the price. This approach also allows students to carry out the inverse process, that is, dividing the selling price (the price including tax) by 1.05 to determine the original price before tax.

Students apply these concepts in Items 7, 8, and 9, with Item 9 providing an opportunity to consolidate this process by calculating sales-tax percent, multiplying rates, original prices, and the actual selling prices for a variety of situations.

Task D

pages 36–37

This challenging task demonstrates that over a two-year period an annual population growth rate of 7% results in a population increase of 14.49%. This is found by calculating 1.07 × 1.07 – 1= 0.1449 = 14.49%. Students should have access to calculators for this task.

In Item 12, beginning with the original population, 12,500, and multiplying by 1.07 ten times gives the approximate population after 10 years. Help students see that pressing = on the calculator ten times after entering 12,500 × 1.07 gives the population after 10 years.

Students should find that the population doubles after about 10 years when the annual growth rate is 7%. If calculators are not capable of this repeating function, students should revert to the paper-and-pencil method.

In Item 13, the population doubles (approximately) when the product of the annual growth rate percent times the number of years equals 70. For example, the expectation is that a population with a growth rate of 5% will double in about 14 years, since $5 \times 14 = 70$. While this task will be challenging for many students, there is much to be gained by having students persist with the task, perhaps working together in small groups, as they work toward the generalization described above. It may be helpful to provide additional factor pairs of 70 to assist students with the generalization (for example, 14×5, 2×35, 4×17.5, 3.5×20, and 1.75×40).

Section 5 Investigating Graphs and Changing Rates pages 38–41

GOALS

To use graphs to represent changes in height of liquid in a container as related to changing volume of liquid added

To predict the shapes of containers, based on the changing heights of liquids represented by lines on graphs

Overview

Students continue to work with graphs that show rates and changes in rates. They also discover how the shape of a container affects the slope of a line that relates the height of liquid in the container to an increasing volume of liquid being added.

Task A pages 38–41

In this task, students predict the rate at which water height changes as water is poured steadily into different-shaped containers. Pouring water steadily means that the volume of water in each container is the same at any moment. In making their predictions, students must first try to make generalizations about the shapes of graphs that illustrate different rates of change. Thinking about graphing in this way should help students better understand the relationship between changing rates and the resulting graphs, a fundamental skill for algebraic thinking.

In Item 1, because the shape of container B is narrower than container A, the height of the water in container B increases at a faster rate than does the water height in container A, even though the volume of water in each container is the same at any moment. The line for container B will be steeper than the line for container A. In the same way, students must reason that the line for container C is less steep than the line for container A. Pouring water steadily into real containers with vertical sides will allow students who have difficulty with these ideas to see how the height changes as the volume increases.

In Item 3, students have another opportunity to work with graphs showing how the height of water changes as the volume increases in different containers. Here, the largest container, container D, is also the tallest container (compare this with container C in Task A). Students sometimes confuse the height of the containers with the rate at which the height of liquid added changes, which is the focus of this task.

In Item 4, container F can be thought of as two joined containers. In the bottom part, the height increases at a faster rate than it does in the upper part. The graph for this container shows the first part representing the bottom part of the cup as steeper than the second part representing the upper part. Students can see that container G also has two parts, so its graph will also have two separate slopes.

In Item 5, each graph shows how the height of water changes as the volume increases in containers that might be thought of as three containers joined together. Students who successfully draw these containers will have demonstrated a clear understanding of the link between the slopes, or steepness, of the graphs and the shapes of the containers.

In Item 6, the changes of height as the volume steadily increases in different containers are represented by lines with an increasing number of segments. Students will find these exercises challenging and will benefit from having an opportunity to work with other students in making decisions as to the shape of the parts of the graphs.

In Item 7, the containers and their graphs from Item 6 provide a clue as to what the graph for each container will look like. These experiences lead students toward making generalizations about the shapes of graphs where one variable (height) is *continuously* changing rather than changing at a steady rate, even though the volume is increasing steadily. Students who have extended opportunities to explain these ideas in their own words initially and to then work in small groups to discuss their findings will achieve greater understanding.

In Exercise 7a, as the volume of water increases steadily, the rate of change of the height of water continuously increases. So the graph becomes increasingly steep.

Finally, in Exercise 7b, as the volume of water increases steadily, the height increases at a gradually slower rate. So the graph becomes less steep.

Annotated Student Pages

Investigating and Interpreting Graphs

Task A

The heights and weights of several children are shown on the graphs below.

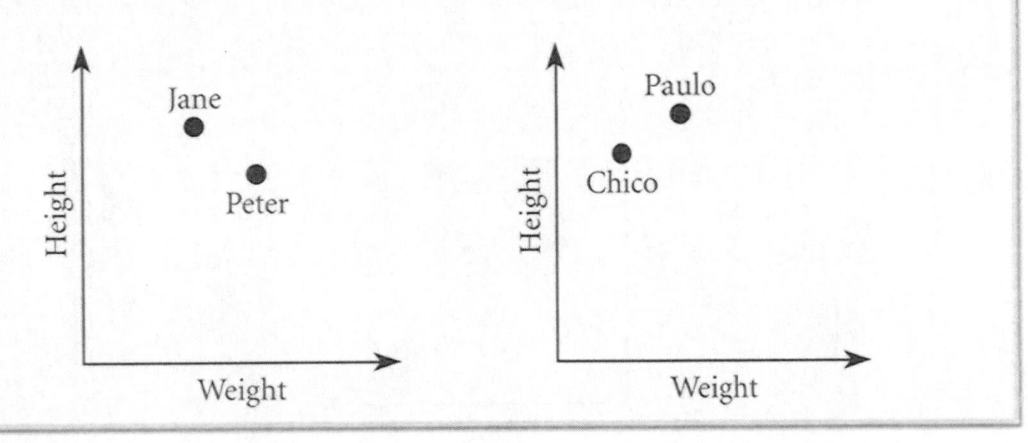

1. Choose from the following word list to complete each sentence below.

lighter	heavier	shorter	taller

a. Peter is heavier than Jane and is also shorter than Jane.

b. Paulo is heavier than Chico and is also taller than Chico.

2. Mark and label points to show Alvin, Kim, and Chuck on the following graph.

Chuck is heavier than Alvin but is lighter than Kim. Alvin is taller than Chuck and Kim. Chuck is shorter than Kim.

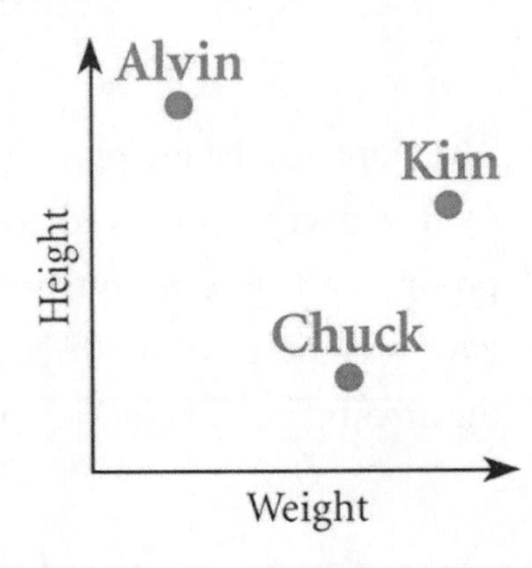

3. Describe the relationship between the points shown on each graph.

a.

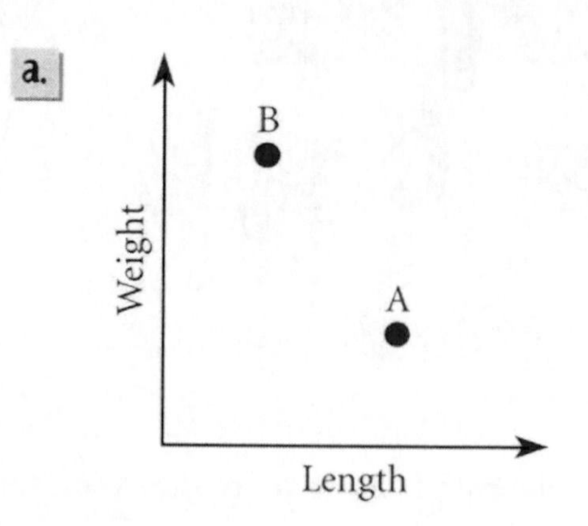

A is longer and lighter than B.

or

B is heavier and shorter than A.

b.

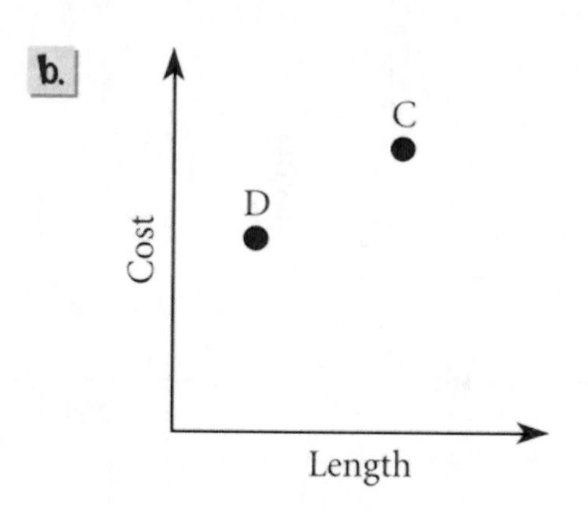

C is longer and more costly than D.

or

D is shorter and less costly than C.

c.

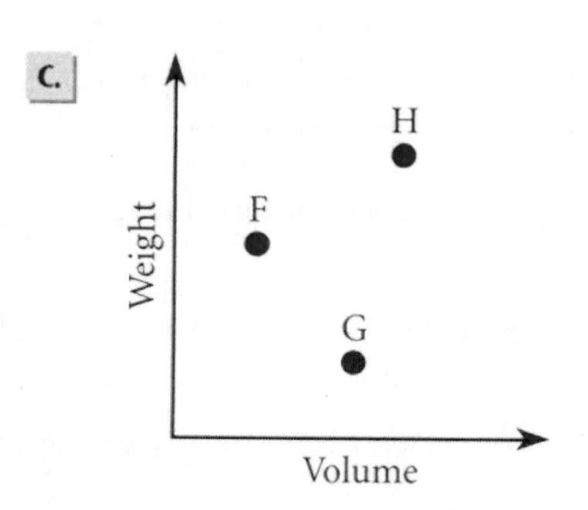

H is the heaviest and has the greatest volume. G is lighter than F and has greater volume than F.

or

F has the least volume and G is the lightest. H is the heaviest and has the greatest volume. F is heavier than G.

4. The graph on the top of page 7 shows the weight and cost for two packages of frozen corn produced by both Maple Foods and Oak Foods. If the products are the same quality, which company offers a better price? Explain your reasoning. (Hint: Think about the relationship between the cost and the amount of corn in each package.)

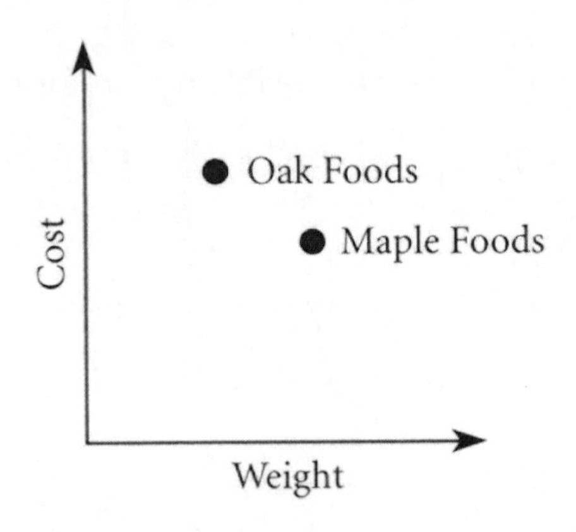

Maple has the better price because its package is heavier and cheaper than Oak's package. This means you get more for less money from Maple.

Task B

Zoe wants to buy canned tomatoes for her pizzeria. Cedar Foods makes cans in three different sizes. They are labeled A, B, and C on the graph. The graph below shows the cost of purchasing 1, 2, 3, 4, and 5 cans for size A, 1 and 2 cans for size B, and 1, 2, and 3 cans for size C.

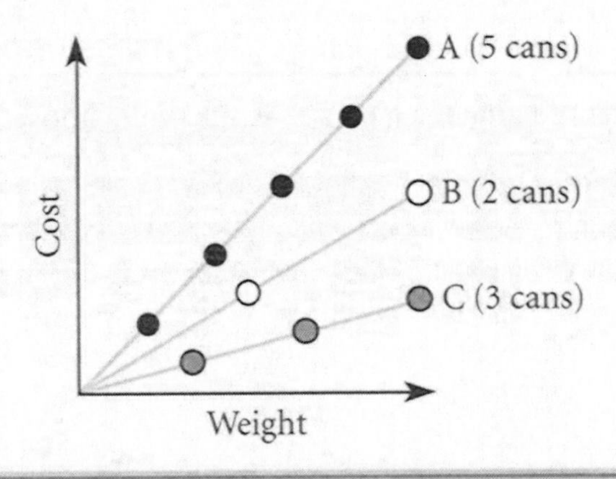

5. Use the graph above to complete each sentence. Explain your reasoning.

   a. The small-size can is labeled A.

   b. The medium-size can is labeled C.

   c. The large-size can is labeled B.

   The weight of 5 size A cans equals the weight of 2 size B cans and 3 size C cans.

6 Zoe decides to buy the size B cans. Give a possible reason for her decision.

The amount of canned tomatoes she needs to prepare her sauce is about equal to the volume of can B. If she uses size A, she will have to open three cans, and some of this will be wasted. If she uses size C, she will have to open two cans, and some will be wasted. In either case she would end up paying more for the tomatoes she uses than if she opens one size B can.

7 Complete the sentence below. Then explain your reasoning.
You get the best price if you buy size __C__.

The weight of five size A cans, two size B cans, and three size C cans is the same. For this weight, size C is cheapest.

8 If you lived alone, which size would you choose? Explain your reasoning.

Any of the cans might be chosen, depending on how much tomato is required. If you needed only the amount of tomato in a size A can, then you might choose size A cans. You might also cook 4 meals at once and freeze 3 of them for use later. Then size C or even size B might be best.

Task C

Zoe says she has an easy way to find the best price among Oak Foods, Maple Foods, and Cedar Foods. She begins by marking points on a graph for identical products that have different weights and prices. See Graph 1.

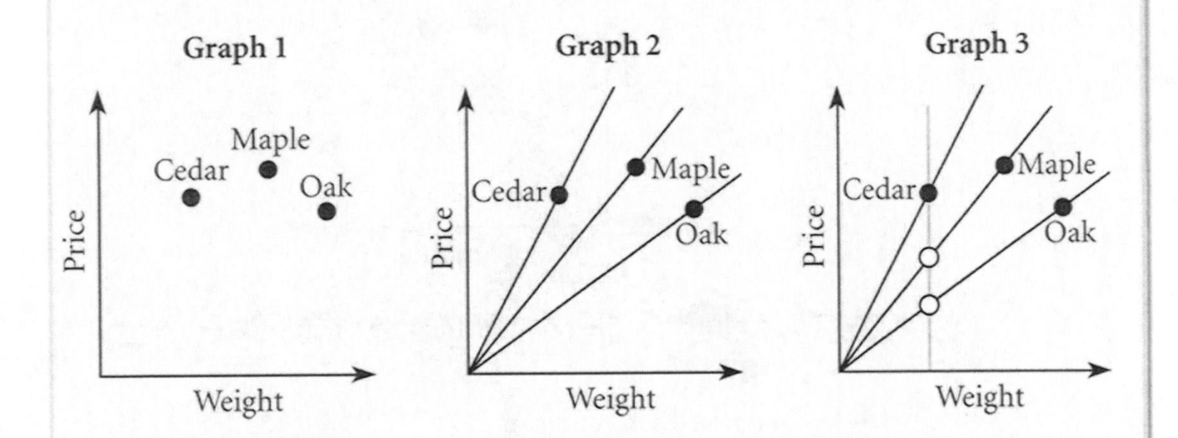

She then draws the three lines shown in Graph 2 and finally adds a vertical line as shown in Graph 3.

Zoe says that the vertical line in Graph 3 shows that Maple offers a better price than Cedar. Explain her reasoning.

All points on any vertical line represent the same weight. Moving down the vertical line shows decreasing price. Cedar is above Maple, which is above Oak. So for an equal weight of product, Cedar costs more than Maple, which in turn costs more than Oak.

10. Complete the following sentence (Hint: Use the vertical line in Graph 3 to help you).
The best price is offered by **Oak**.

11 The following graph shows the prices for three different amounts of cooking oil sold by three different companies.

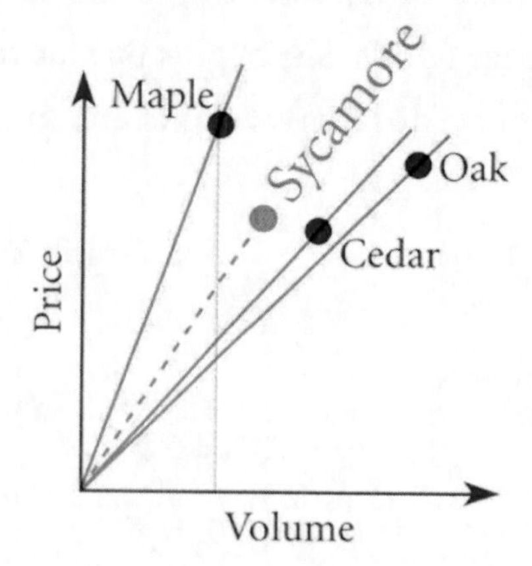

a. Draw lines on the graph using Zoe's method.

b. Complete each sentence.

1. The best price is offered by <u>Oak</u>.
2. The worst price is offered by <u>Maple</u>.
3. Sycamore's prices for cooking oil are better than Maple's prices but not as good as Cedar's. On the graph in Item 11, mark and label a point for an amount of oil from Sycamore.

12 Study the graph below, then explain how the gray horizontal line shows which of the companies, A, B, or C, has the best price for oil.

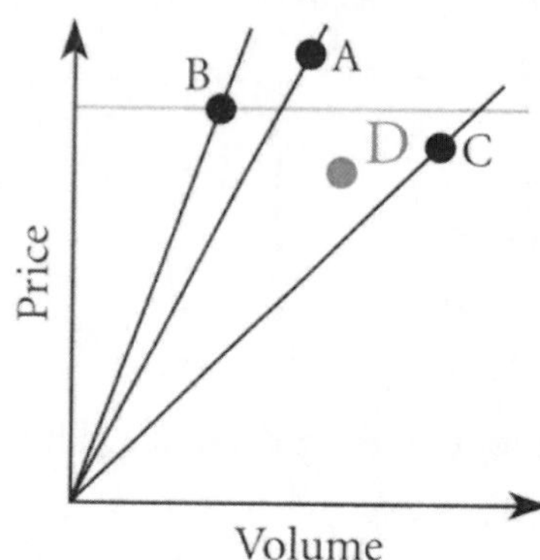

All points along the gray horizontal line represent the same cost. Moving along the line to the right represents increasing volume of oil for the same price. So, A is better than B, and C is better than A and B. So, C has the best price.

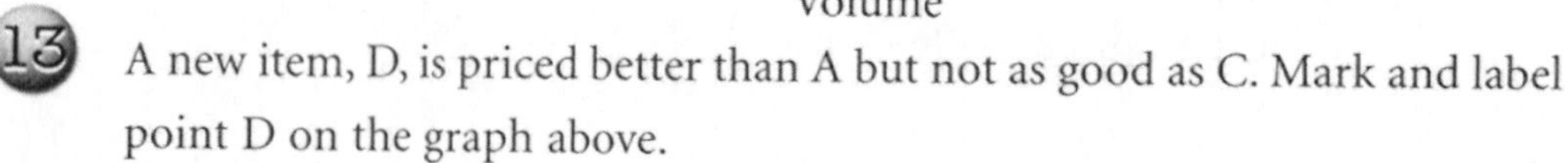

13 A new item, D, is priced better than A but not as good as C. Mark and label point D on the graph above.

Page 42

Task D

Carl, Donovan, and Michael run a 200-meter race. On the graph below, C stands for Carl, D stands for Donovan, and M stands for Michael. The graph shows that Michael is the winner and Donovan comes in second.

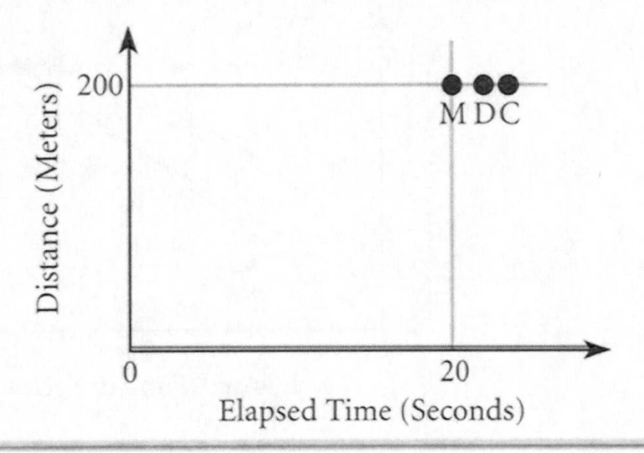

14 Explain how the three points on the graph show who comes in first, second, and third.

All the points along the gray horizontal line represent the same distance, 200 m. The order of the points along the line reflects the order in which they finished. Michael was the winner so his point comes first, followed by the points for Donovan and Carl.

15 On the graph below, lines are drawn from each point to the origin (0, 0) to determine who was leading at 10 seconds into the race.

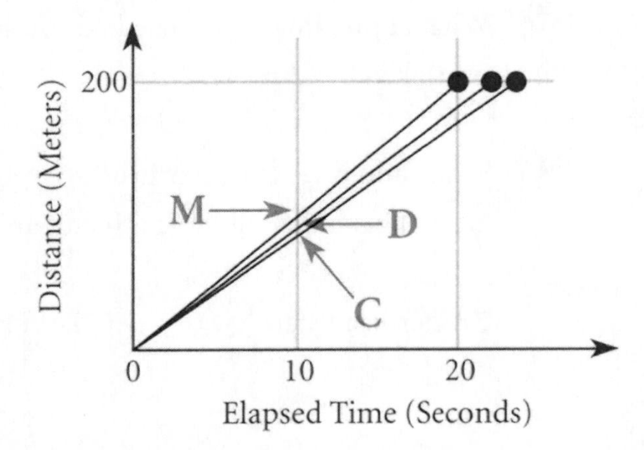

On the graph, find the vertical line that represents 10 seconds. Mark and label points C, D, and M to show the likely positions of each athlete at 10 seconds into the race.

16 This graph shows what happened when Janine (J), Chrissy (C), and Annie (A) competed in a 50-meter race. Janine and Annie had a head start. Use the graph to answer each question below. (Hint: Notice that the starting point for the race is different for each runner.)

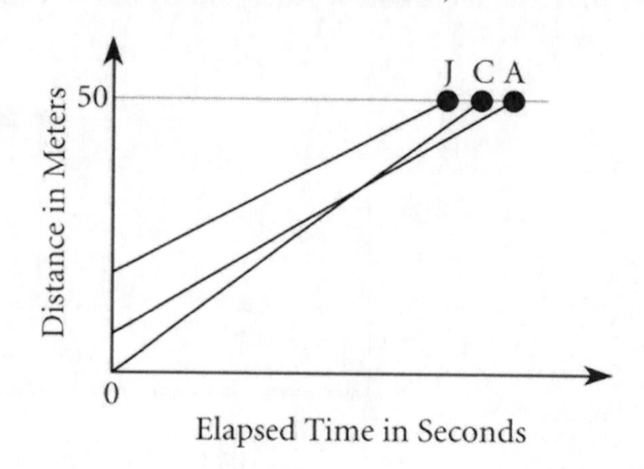

a. Who won the race? J

b. Who ran the farthest? C

c. Who was the fastest? C
(The race line slope, or rate, is the steepest.)

d. Who was the slowest? J
(The race line slope, or rate, is the least.)

e. What approximate distance did Janine run? about 30 m

f. What approximate distance had Annie run before she was overtaken by Chrissy? about 35 meters

g. Who should get a greater head start and who should get a lesser head start so that all three athletes are likely to finish together?

Greater head start A Lesser head start J

17 On the following graph, the lines representing Annie's run and Janine's run have been extended on both ends so they can be adjusted up and down as well as left and right. Draw new lines for Annie and Janine to show the correct amount of head start for both of them so that all three athletes are likely to finish the race together.

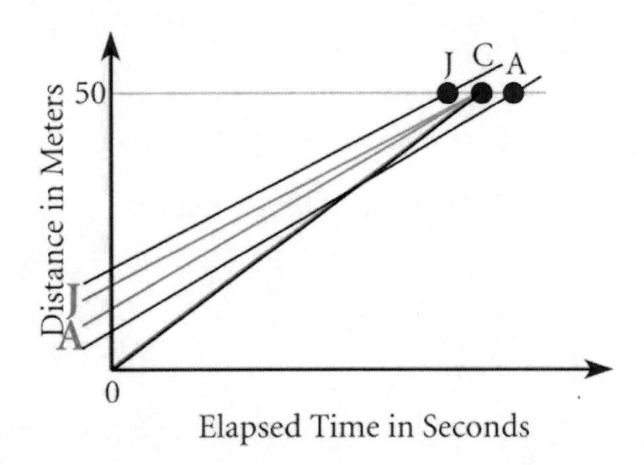

18 Hamish (H) and Scott (S) compete in a triathlon. They run, cycle, and swim a total of 30 kilometers each.

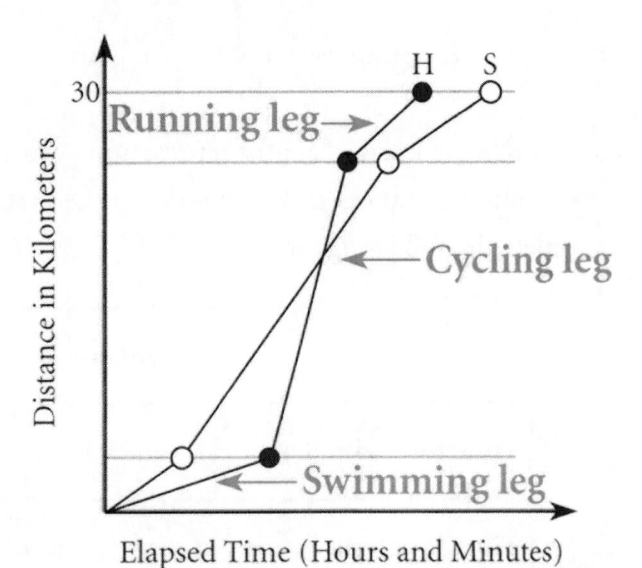

a. Label the parts of the graph that represent the running, cycling, and swimming segments of the race. (Hint: Use what you know about slope and how the steepness of the line relates to the rate or to the speed for each sport.)

b. Who is the faster runner?

Hamish (Hamish's running line is steeper than Scott's running line. This means that for any elapsed time, Hamish's line rises more than Scott's line, that is, Hamish runs farther than Scott in any given time period. So Hamish must be running faster than Scott.)

c. Who is the slower cyclist?

Scott (Hamish's cycling line is steeper, so he is cycling faster than Scott.)

d. Who is the faster swimmer?

Scott (Scott's swimming line is steeper, so he is swimming faster than Hamish.)

e. What is the approximate distance for each leg of the triathlon?

Running leg = about 5 km Cycling leg = about 20 km

Swimming leg = about 4 km

19 Use a ruler to measure lengths representing elapsed times as well as lengths representing distances in kilometers on the graph in Item 18 to help you complete the following table.

		Hamish	Scott
a.	**Time for Swim**	About 1 h	About 30 min
b.	**Time for Cycle**	About 30 min	About 1 h 10 min
c.	**Time for Run**	About 35 min	About 45 min
d.	**Total Time**	About 2 h 5 min	About 2 h 25 min

e. About what time into the triathlon did Hamish overtake Scott?

about 1 h 25 min

The following graph shows the depth of water in a bathtub as Melanie prepares and then takes a bath.

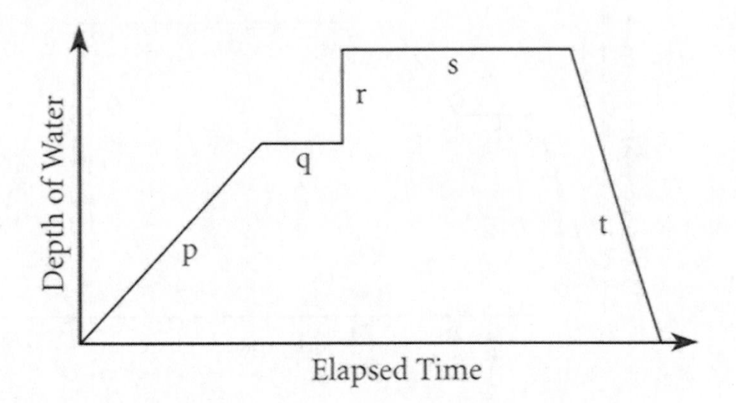

She writes this sentence to describe what the segment p stands for:

Melanie fills the bathtub with water.

20 Write a sentence for each segment on the graph to describe what Melanie might be doing as she prepares and takes her bath. The first one is done for you.

a. segment q *Melanie turns off the bath water and waits a little while before she gets in.*

Possible answers:

b. segment r Melanie gets into the bath, and the water level rises suddenly.

c. segment s Melanie is sitting or lying down in the bath.

d. segment t The bath is emptying faster than it was filled.

21 Each day Alvin walks to and from school. The following graph shows how far Alvin is from home at any moment from the time he leaves home in the morning until he returns in the afternoon.

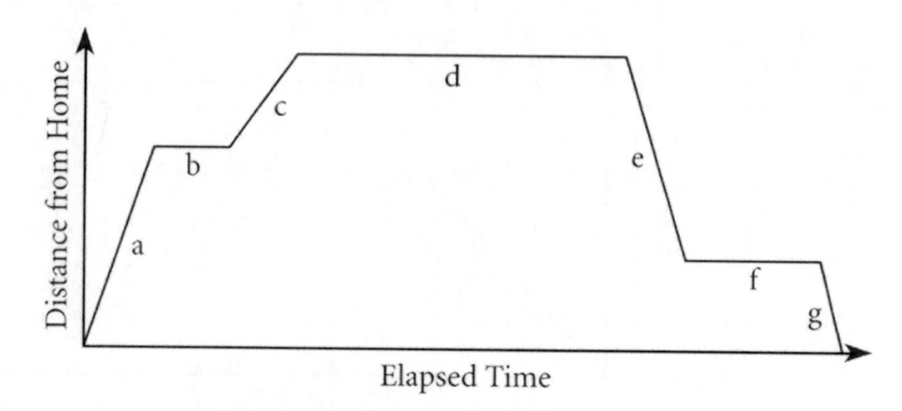

For each segment in the graph above, write a sentence that tells you what Alvin might be doing on his way home.

Possible answers:

a. segment a Alvin is walking at a constant speed.

b. segment b Alvin stops at a friend's home on the way to school.

c. segment c Alvin and his friend walk to school, but not as fast as Alvin walked to his friend's home.

d. segment d Alvin is at school.

e. segment e Alvin and his friend run to his friend's home.

f. segment f Alvin plays at his friend's home.

g. segment g Alvin walks quickly to his home.

Mike goes to see his friend Billy, who lives in the neighborhood. They talk for a short while and then Mike returns home. One of the following graphs shows how the distance Mike is from home changes during that time.

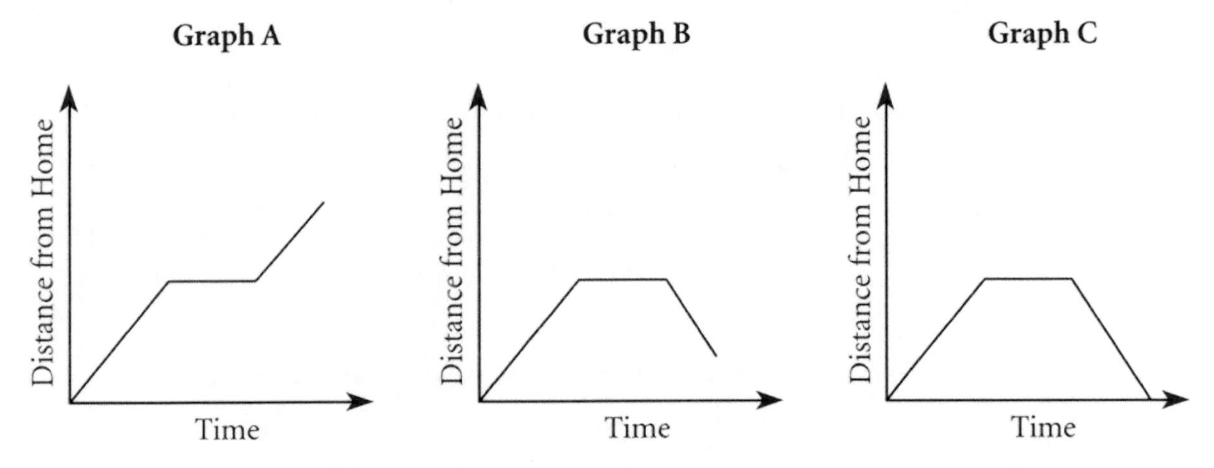

a. Which graph best shows how the distance Mike is from home changes when he goes to see his friend?

Graph C

b. Complete each sentence to explain why you did not choose each of the other two graphs.

(Possible explanations are given)

1. I did not choose Graph A because it shows Mike goes to his friend's home, talks for a while, then walks farther away from his home.
2. I did not choose Graph B because the end of the graph shows that, although Mike began the journey home, he did not get there.

Kelly says that the following graphs represent trips she has made.

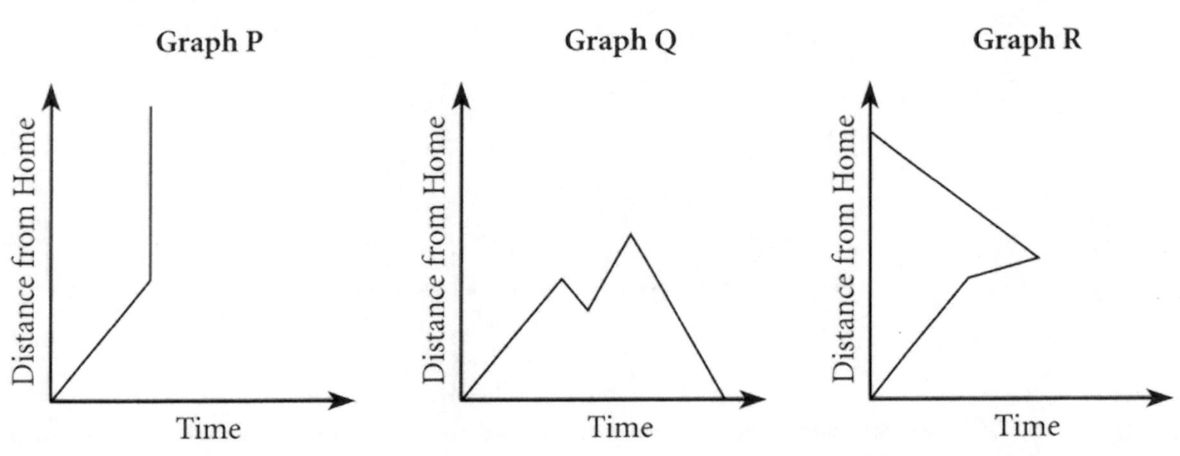

a. Decide whether Graph P on page 17 can represent a trip. Explain your reasoning.

Graph P can never represent a trip. The vertical part of the graph indicates that Kelly is at an infinite number of distances from home at the same time. This is impossible.

b. Decide whether Graph Q on page 17 can represent a trip. Explain your reasoning.

Graph Q can represent a trip. It shows Kelly went away from home, then went toward home, then went away from home, and finally went home.

c. Decide whether Graph R on page 17 can represent a trip. Explain your reasoning.

Graph R can never represent a trip. Many vertical lines drawn on the graph would cross Kelly's trip line at two places. This would mean she was in two places at the same time. This is impossible.

Larry wants to meet a friend who lives 200 miles away. He and his friend decide to meet at a restaurant that is on the road between their homes.

The graph below shows parts of the trip made by Larry and his friend.

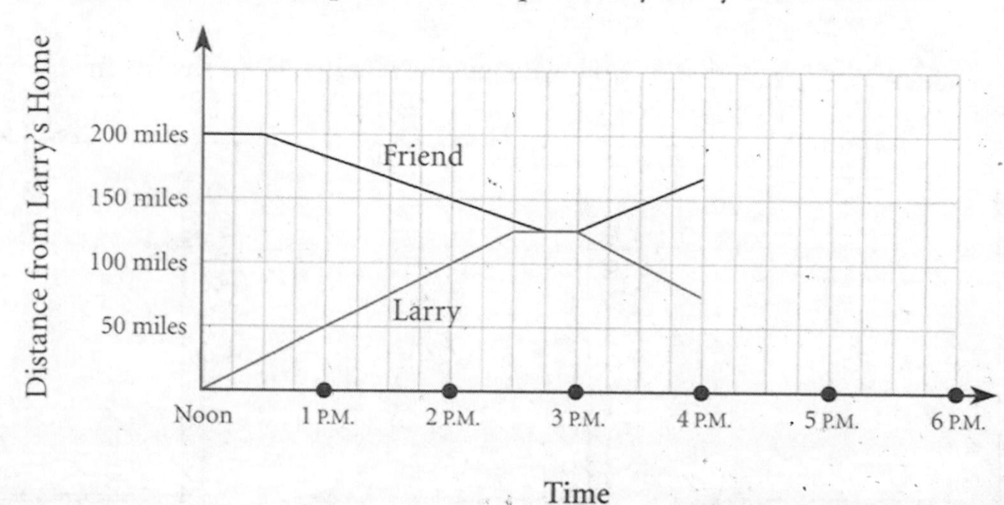

24 Larry leaves his home at noon. He arrives at the restaurant at 2:30 P.M.

a. How far is the restaurant from Larry's home?

about 125 miles.

b. How long did Larry wait before his friend arrived?

about $\frac{1}{4}$ hour (15 min)

c. How long did Larry wait at the restaurant before he began the trip home? about another $\frac{1}{4}$ hour (15 min). He was at the restaurant for about 30 minutes total.

d. How far did the each person travel to reach the restaurant?

Larry traveled about 125 miles and his friend traveled about 75 miles.

e. At what time did Larry's friend start his drive toward the restaurant? 12:30 P.M.

f. How long did Larry's friend take to reach the restaurant?

about 2 hours 15 minutes.

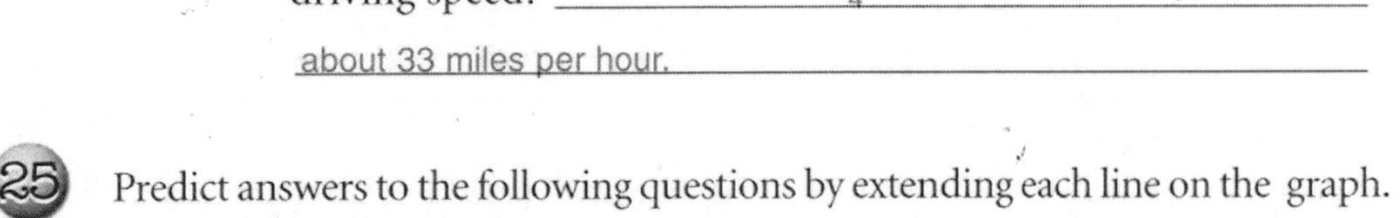

g. Larry drove at a steady 50 miles per hour. What was his friend's driving speed? about 75 miles in $2\frac{1}{4}$ hours, that is, at a speed of about 33 miles per hour.

25 Predict answers to the following questions by extending each line on the graph.

a. When did Larry arrive back at his home? about 5:15 P.M.

b. When did Larry's friend arrive back at his home?

about 5:00 P.M.

Investigating Measurement Conversion Rates

Task A

Marge wants to keep track of travel distances in miles as well as kilometers. She made this graph to help her convert miles to kilometers as well as kilometers to miles.

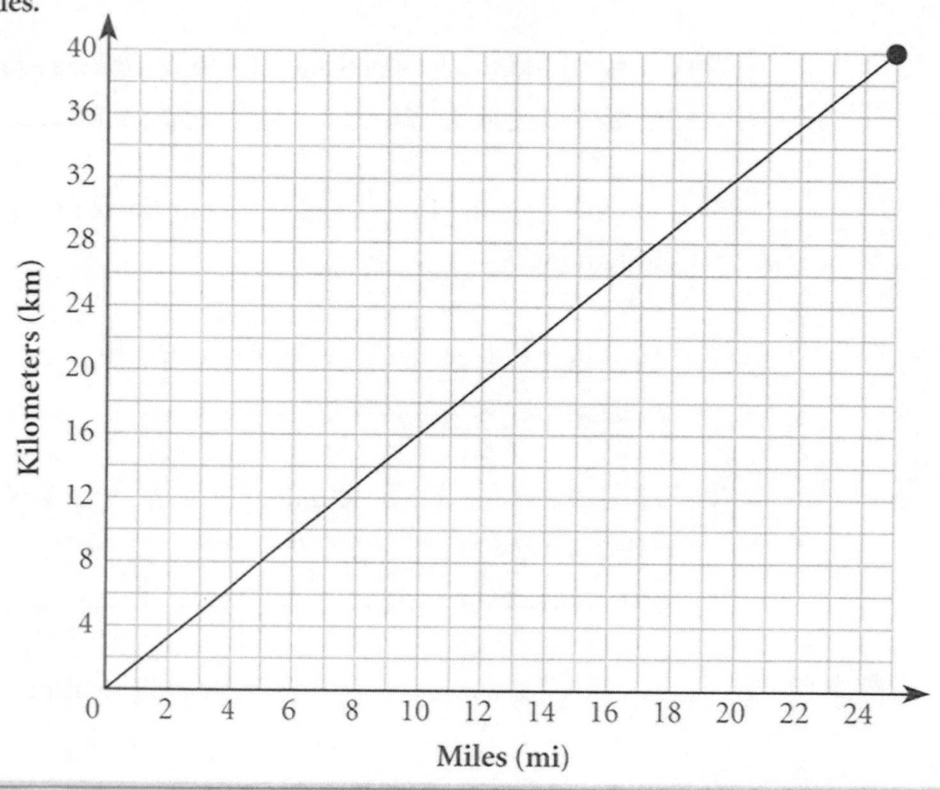

Use the graph and patterns to help you complete the following table.

mi	km
25	40
20	32
15	24
10	16
1	1.6
95	152

2. Complete the diagram below. In the top box, write the number you can multiply by to convert miles to kilometers. You can divide by the same number to convert kilometers to miles. (Hint: Use your answers from the table in Item 1 to help you.)

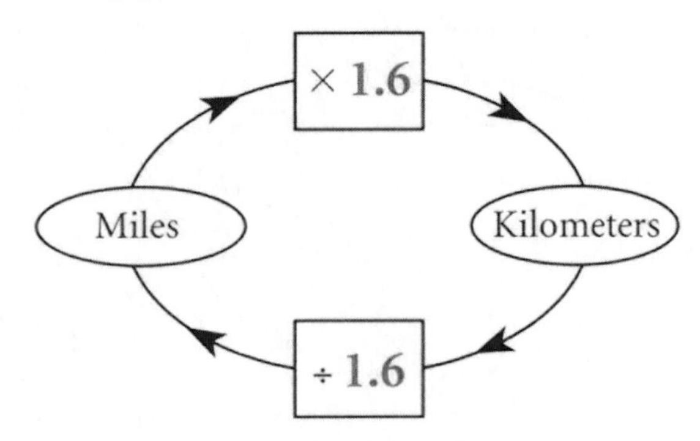

3. Use the conversion diagram above to complete the following tables.

a.

mi	km
43	**68.8**
625	1,000
250	**400**

b.

mi	km
26.2	**41.9**
2.5	4
10,000	**16,000**

c.

mi	km
11.1	17.83
37 .6	**60.2**
0.625	1

4. Marge uses division to convert kilometers to miles. Marge's friend Paul says he can use multiplication to convert kilometers to miles. Complete the diagram below. (Hint: Use your answers from the three tables above to help you.)

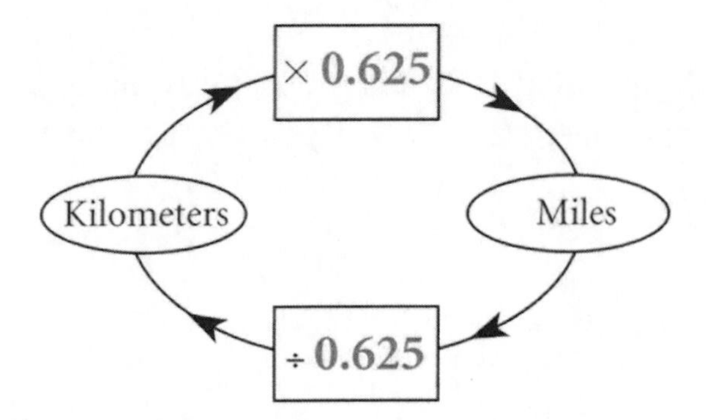

5. Which is the faster speed, 100 kilometers per hour (km/h) or 65 miles per hour (mph)? Explain your reasoning.

The faster speed is 65 mph. Since 65 miles = 104 kilometers, 65 mph is faster than 100 km/h.

Task B

Josh wonders how accurate the graph is for converting miles to kilometers. He knows that 1 inch equals 2.54 centimeters and that there are 63,360 inches in a mile.

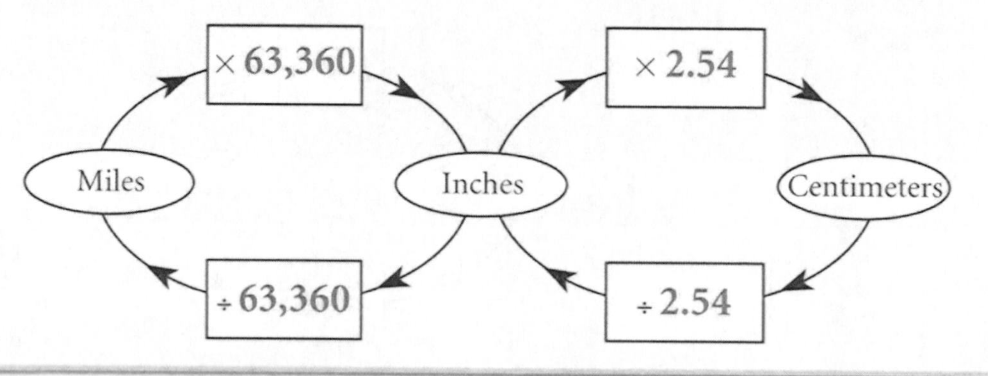

6. Use Josh's information to complete the conversion diagram in Task Box B.

7. Use the conversion diagram above to help you complete the following table. (Hint: 100 centimeters = 1 meter, and 1,000 meters = 1 kilometer)

Miles	Centimeters	Meters	Kilometers
1	160,934.4	1,609.344	**1.6093**
10	**1,609,344**	**16,093.44**	**16.093**
0.621	**100,000**	1,000	1
6.214	1, 000,000	**10,000**	**10**

8. A marathon is 26 miles 385 yards, or 26.21875 miles. How far is it in kilometers?

42.195 km

You can use this graph to convert kilograms to pounds or pounds to kilograms.

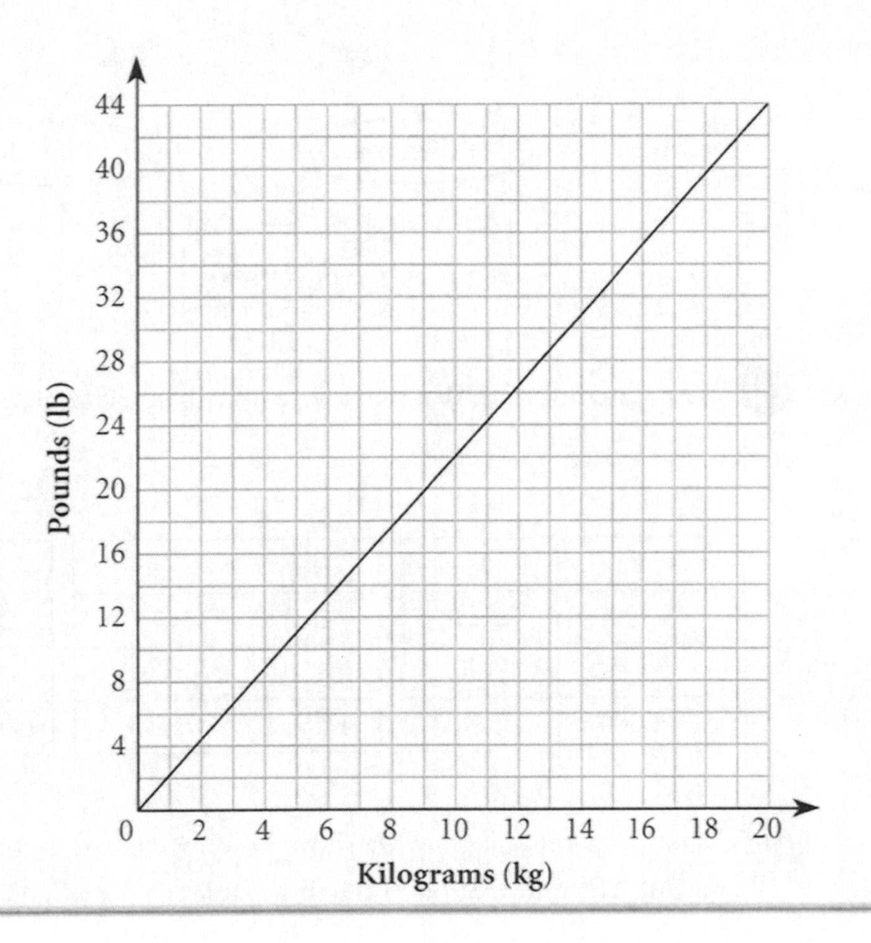

9 Use the graph and look for patterns to complete the following table.

Kilograms	Pounds
20	44
10	22
1	2.2
70	154

10 Use the information in the table in Item 9 to help you complete the following conversion diagram.

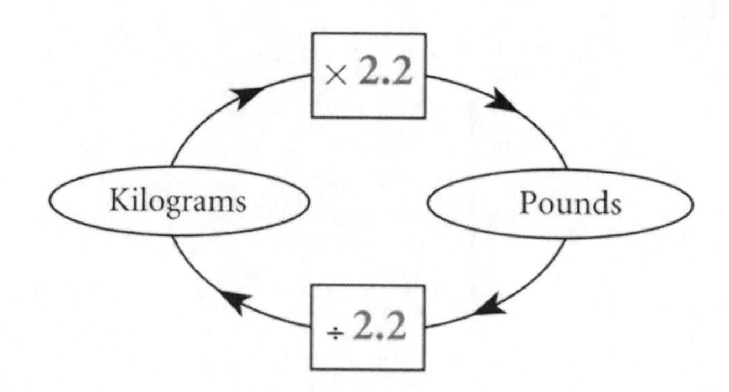

11 Use the conversion diagram above to help you complete each table.

a.

kg	lb
100	220
45.45	100
1,000	2,200

b.

kg	lb
6.4	14
150	330
109.1	240

c.

kg	lb
50.9	112
750	1,650
0.4545	1

12 Complete this diagram to show how to convert pounds to kilograms. (Hint: The information in the three tables above will help you.)

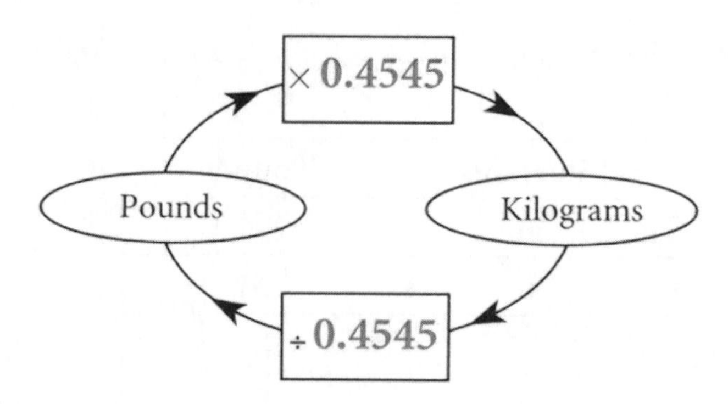

You can use this diagram to convert speed in miles per hour to speed in yards per second.

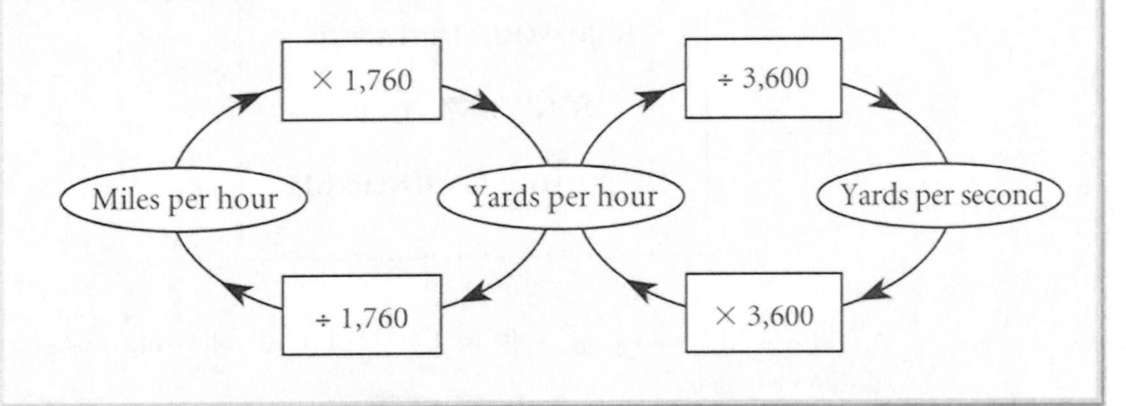

Explain why the numbers 1,760 and 3,600 have been used in the boxes.

1 mile = 1,760 yards. So, the number of yards traveled in 1 hour = 1,760 × number of miles traveled in 1 hour. The speed in yards per hour = 1,760 × speed in miles per hour. There are 60 × 60 = 3,600 seconds in 1 hour.

So the distance traveled in 1 second = $\frac{1}{3,600}$ of the distance traveled in 1 hour, that is, speed in yards per second = speed in yards per hour ÷ 3,600.

Use the speed-conversion diagram above to help you complete the following table.

Speed (miles per hour)	Speed (yards per second)
60	**29.33**
61.36	30
204.5	100
2	**0.98**

Carla knows that to avoid tailgating collisions, motorists are advised to keep at least 2 seconds driving distance behind the vehicle in front.

KEEP YOUR DISTANCE

DON'T BE A FOOL

OBEY THE 2-SECOND RULE

Complete the following table to find 2-second following distances for different speeds.

Speed (miles per hour)	Speed (yards per second)	2-Second Following Distance (yards)
64	31.3	62.6 or about 63
51	**24.9**	**49.9** or about **50**
37	**18.1**	**36.2** or about **36**
28	**13.7**	**27.4** or about **27**
2	**1.0**	**1.96** or about **2**

Carla found a simple rule that is easy to remember and that helps drivers choose a safe following distance. Write the rule. (Hint: Look carefully at the numbers in the first and third columns in the table above.)

A very good rule is that the safe following distance in yards equals the speed in miles per hour. So the safe following distance when traveling at 60 mph is 60 yards.

Page 43

Investigating Exchange Rates

When Carole and Bennie return from a backpacking vacation in New Zealand, they each have some New Zealand dollars that they want to change into United States dollars. The following graph shows equivalent amounts in two currencies.

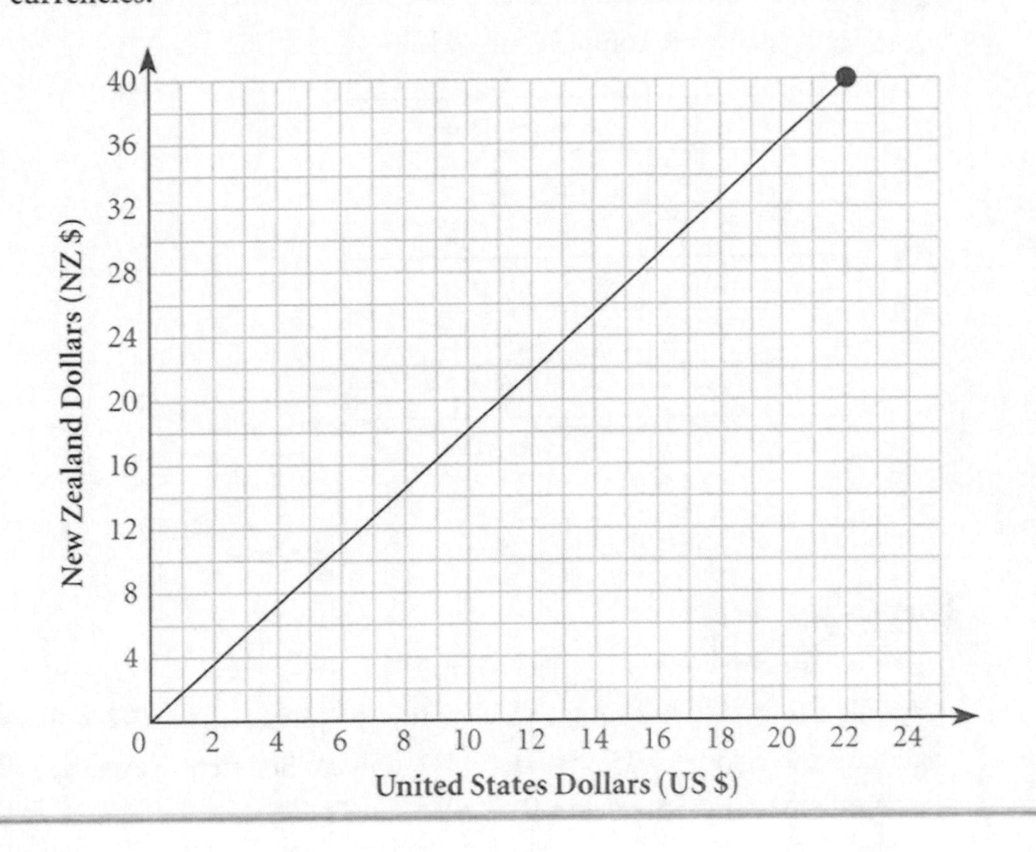

1. Carole has NZ$30. How many US dollars should she expect to get? ________
   between US$16 and US$17

2. Bennie has NZ$14. How many US dollars should he expect to get? ________
   just under US$8

3. Use the graph from page 27 to help you complete each table. (Round your answers to the nearest dollar.)

a.

US$	NZ$
6	**11**
8	15
14	**26**

b.

US$	NZ$
18	32
3	**6**
12	**22**

c.

US$	NZ$
18	**33**
19	34
2	**4**

4. Carole notices from the graph that NZ$40 = US$22. Use this information and patterns to complete the table.

NZ$	US$
40	22
20	**11**
10	**5.5**
1	**0.55**
87	**47.85**

Task B

Carole thinks she has found an easy way to convert from NZ dollars to US dollars and also from US dollars to NZ dollars. She first draws the following diagram for converting from NZ dollars to US dollars.

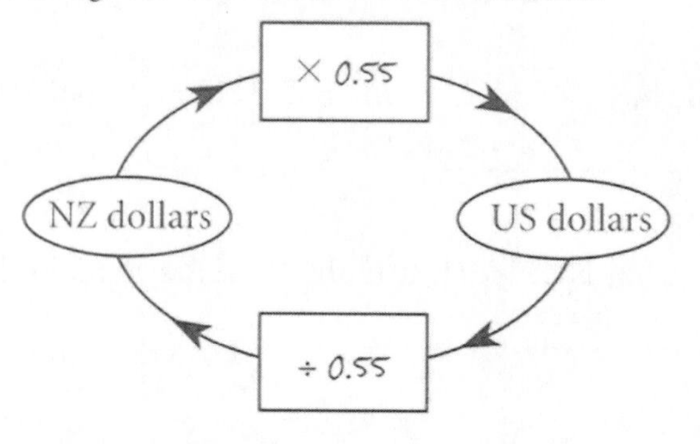

Then she writes the same number in both boxes. This number is the exchange rate for converting NZ dollars to US dollars.

5. Complete the following. (Hint: The information in the table in Item 4 will help you.)

NZ$1 = US$ **0.55**

The exchange rate for converting NZ dollars to US dollars is **0.55**.

6. Carole divides both sides of the rule in Item 5 by 0.55 to write an equation showing how many NZ dollars equal one US dollar. Complete the following.

US$1 = NZ$ **1.82**

The exchange rate for converting US dollars to NZ dollars is **1.82**.

Task C

Toshi wants to send 20,000 yen (¥20,000) to a mail-order company in Tokyo to buy a camera. He uses the following diagram to help him calculate how much the camera will cost in US dollars.

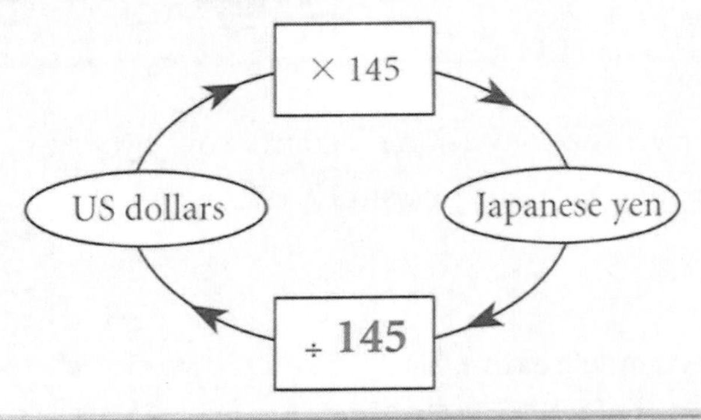

7. Complete the conversion diagram above by writing the number that goes in the ÷ box.

8. How many US dollars does Toshi need for his camera? **$137.93**

9 Complete the following tables.

a.

US$	¥
10	1,450
6.90	1,000
100	14,500

b.

US$	¥
20	2,900
31.03	4,500
500	72,500

c.

US$	¥
275.86	40,000
37	5,365
689.66	100,000

Task D

Each month Julio sends US$85 home to his parents in Guerrero Negro, Mexico.

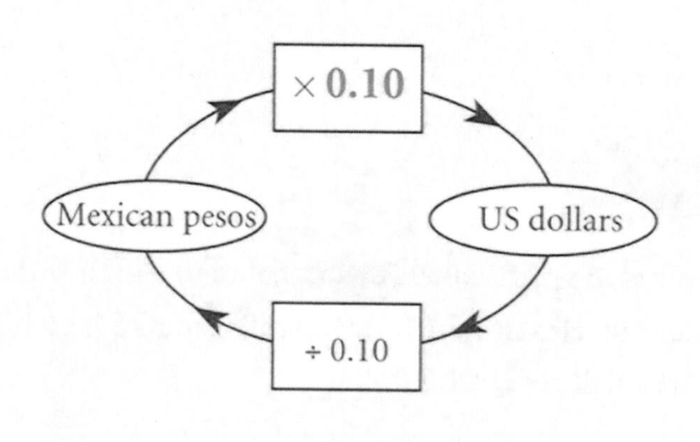

10 Complete the conversion diagram above by writing the exchange rate for converting Mexican pesos to US dollars.

11 Now complete each table.

a.

Pesos	US$
10	1
10,000	1,000
100	10

b.

Pesos	US$
25	2.50
4,530	453
262	26.20

c.

Pesos	US$
50,550	5,055
37	3.70
100,000	10,000

Carole and Bennie planned to stop in Australia on their way home from New Zealand. They used the following currency conversion diagram to help them calculate how much money they would need for the stopover.

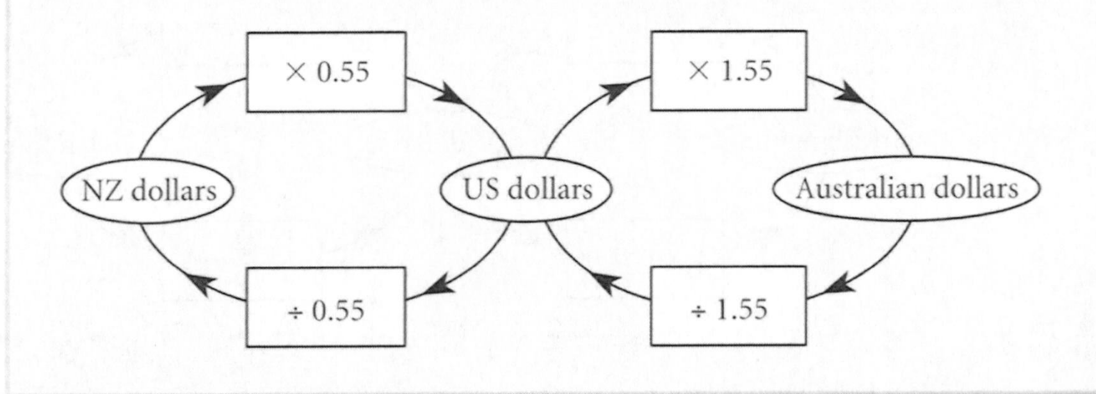

12 Complete the following exchange rate.

US$1 = **1.55** Australian dollars (A$)

13 Carole and Bennie saved US$5,000 for their vacation. Suppose they spend all of this in New Zealand. How much will they spend in NZ dollars?

$9,091 ($5,000 ÷ 0.55 = $9,091)

14 One alternative they considered was spending US$3,500 in New Zealand and the remainder in Australia. Complete each equation.

a. US$3,500 = NZ$ **6,364** **b.** US$1,500 = A$ **2,325**

15 When they were in New Zealand, they decided to change NZ$1,000 into Australian dollars. How much will they get in A$? Show your work.

NZ$1,000 = A$ **852.50**

NZ$1,000 × 0.55 = US$550; US$550 × 1.55 = A$852.50

16 Complete the exchange rate for converting NZ$ to A$.

NZ$1 = A$**0.8525** (Since 0.55 × 1.55 = 0.8525.)

Task F

This diagram shows currency conversion for British pounds, US dollars, and French francs.

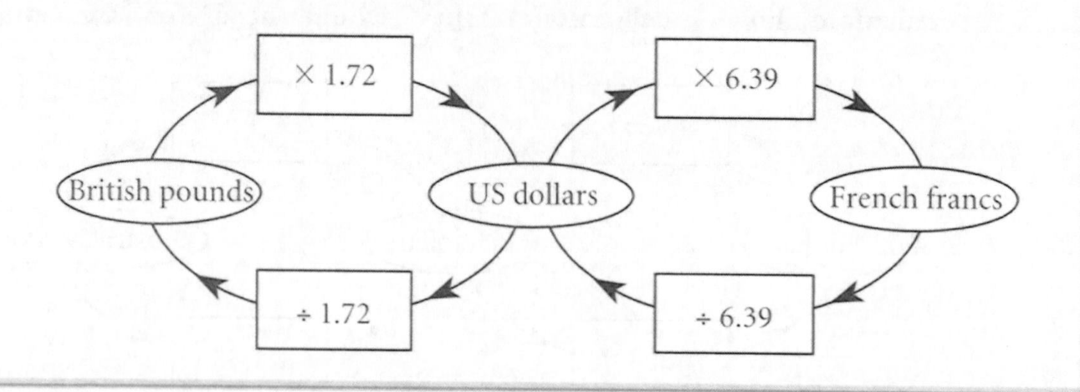

17 Use the conversion diagram to complete the table.

British Pounds Sterling (£)	United States Dollars ($)	French Francs (FF)
1,000	**1,720**	**10,990.80**
581.40	1,000	**6,390**
90.99	**156.49**	1,000

18 Use the table in Item 17 to complete each currency exchange rate.

a. £1 = FF **10.9908**

b. FF1 = £ **0.09099**

19 Jimmy claims that he knows a shortcut for converting British pounds into French francs. He writes, British pounds × 1.72 × 6.39 = French francs. Explain Jimmy's reasoning.

Jimmy starts with British pounds on the chart. Multiplying by 1.72 converts British pounds to US dollars, and multiplying this by 6.39 converts US dollars to French francs. So multiplying by 1.72 x 6.39 converts British pounds to French francs.

Page 43

Investigating Percentage Rates

Clark's Department Store is advertising these discounts for its annual January Sale.

> Clark's
> Giant Discounts
>
> **SALE**
>
> - Clothing 15% OFF
> - Televisions reduced by 18%
> - All outdoor equipment 12% OFF

Bela wants a new tent that usually sells for $85. She makes the following calculation to find the sale price.

10% of $85 is $8.50.

1% of $85 is $0.85.

So, 12% of $85 is $8.50 + $0.85 + $0.85 = $10.20.

1. How much will Bela pay for the tent? **$74.80**

2. Use Bela's method to find the sale price of a 13-inch television set that usually sells for $112.

   1% of $112 is $1.20. Therefore, 11.20 + 8 × 1.12 = 11.20 + 8.96 = 20.16.
   So, Bela pays $112 – 20.16 = $91.84 for the television set.

Task B

Bela's friend Simone says she has a shortcut for finding the sale price for the tent. She tells Bela that a discount of 12% means that she will pay 88% of $85. She then uses a diagram to help with her calculation.

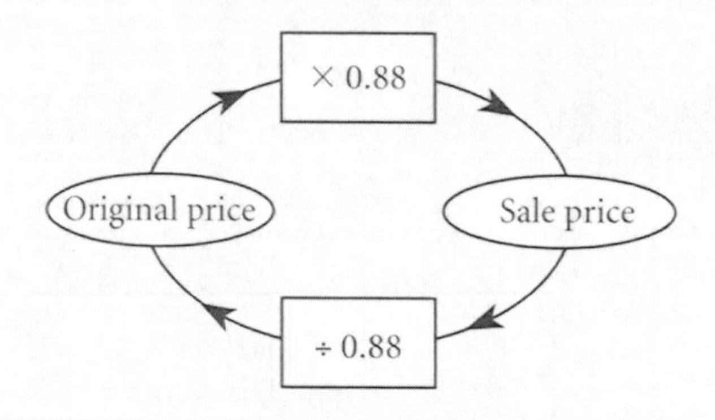

3. Use the diagram and the information in Task Box B to see if you get the same answer for the price of the tent in Item 1. Do the methods agree? **yes**

Students should find that both methods result in the same answer.

4. Explain why Simone multiplies the original price for outdoor equipment by the rate 0.88 to get the sale price.

Imagine cutting the original price into 100 equal parts. A 12% discount means you get to keep 12 of these 100 parts. So, 88 of the 100 parts is paid to the shop. So the shop receives 88 hundredths of the original price, or 0.88 × original price.

5. Use Simone's shortcut to find the original price of a camping stove that has a sale price of $53. Show your work.

Original price = **$60.23**

Original price = $53 ÷ 0.88 = $60.23 (nearest cent)

6. Use Simone's diagram to complete the following chart. Round prices to the nearest 10 cents.

Original Price	Percent of Discount	Multiplying Rate	Sale Price
$36	17	0.83	**$29.90**
$58	**25**	0.75	**$43.50**
$102.40	16	**0.84**	$86
$145.30	**5**	0.95	$138
$86.20	37	**0.63**	**$54.30**
$200	**50**	0.50	$100

Task C

Sandy has to pay 5% sales tax on items she purchases. She has a simple way to work out how much items cost after sales tax is added.

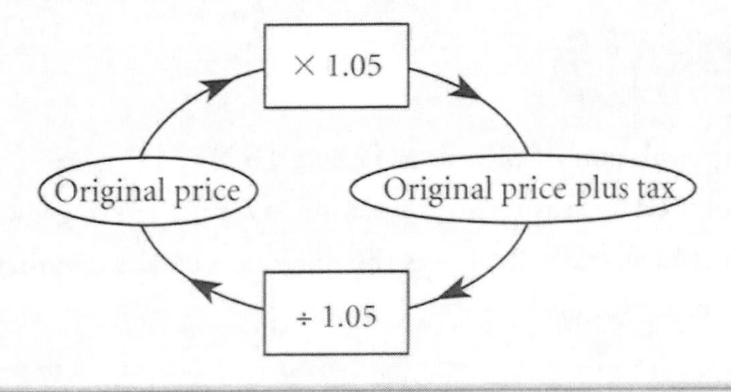

7. Explain why Sandy multiplies the original price by the rate of 1.05 to find the total amount she has to pay.

Imagine cutting the original price into 100 equal amounts. A 5% retail tax means 5 of these amounts have to be added to the original 100 amounts to get 105. So, 105% of the original price = 1.05 × original price.

8. Use Sandy's diagram to find the original price of a pair of running shoes that cost $84.95 including tax. Show your work.

Original price = **$80.90**

Original price + tax = $84.95. Original price = $84.95 ÷ 1.05 = $80.90 (to the nearest 10¢).

Use Sandy's diagram to complete the chart. Round prices to the nearest 10 cents.

Original Price	Sales Tax	Multiplying Rate	Original Price Plus Tax
$36	5%	1.05	**$37.80**
$97	**4%**	1.04	**$100.90**
$46.20	6%	**1.06**	$49
$134.60	**2.5%**	1.025	$138
$86.45	7.75%	**1.0775**	**$93.10**

Page 44

Task D

The population of Silvero is 12,500. Finding oil near the town has led to a population increase. The population is forecasted to grow by 7% per year for the next ten years. Paul uses the diagram below to predict the population of Silvero in two years.

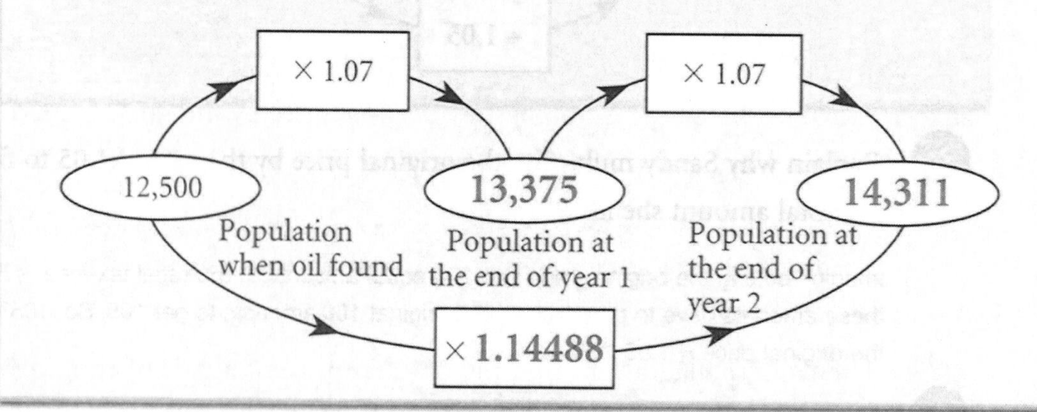

Find the missing numbers to complete the diagram.

Look at Task Box D. Explain why the percentage of population growth over two years is not 14%.

The population increase in the first year is 7% of 12,500, or 875. In the second year it is 936 (7% of 13,375). So, the population increase over the first two years is 875 + 936 = 1,811. This increase is 14.488% of 12,500, not 14%.

Pressing 1 2 5 0 0 × 1 . 0 7 = = on many calculators gives the population after two years. Pressing = repeatedly multiplies the result by 1.07.

a. If your calculator works like this, continue pressing = until you find the population after ten years. about 24,600

b. About how many years does it take before the population doubles? a little more than 10 years

Complete the table to show the number of years it would take the population to double at three different rates of population growth.

Annual Rate of Population Increase	Number of Years for Population of 12,500 to Double
5%	about 14
7%	a little more than 10
10%	about 7

The town planner for Silvero uses the table in Item 13 to make predictions about the town's population.

- **A 14% annual rate of increase will double the population in about 5 years.**
- **A 2% annual rate of increase will double the population in about 35 years.**

Explain how the town planner is able to make these predictions.

The town planner sees that the annual rate of population growth times the number of years for the population to double is about 70. So he calculates 70 ÷ 14 = 5 years, and 70 ÷ 2 = 35 years.

Investigating Graphs and Changing Rates

Task A

Water is steadily poured at the same rate into each of the three containers shown below. Each one will contain the same amount of water at any given time. The graph shows how the height of water in container A changes as the water is poured.

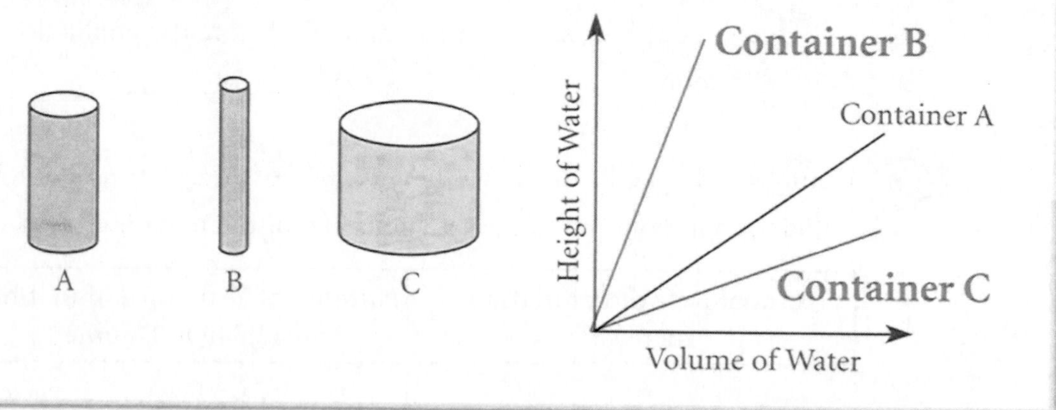

1. On the graph, draw lines for container B and container C to show how the height of the water in container B and container C changes as the volume increases. (Hint: Think about how the shape of the container will affect the height of the water.)

   Note: Graphs shown are approximations. Student representations will vary.

2. Explain why you drew the lines the way you did.

   Imagine pouring liquid into all containers at the same rate. The base of container B is much narrower than the base of container A. As liquid is poured into container B, the level (height) rises more quickly than for container A. So, the graph for B is steeper than the graph for A. The base of container C is wider than the base for container A. As liquid is poured into container C, the level rises more slowly than that of container A. So, the graph for C is less steep than the graph for A.

3. Study the graph below. Then draw lines to show the changes in water height as the volume of water increases in containers D and E.

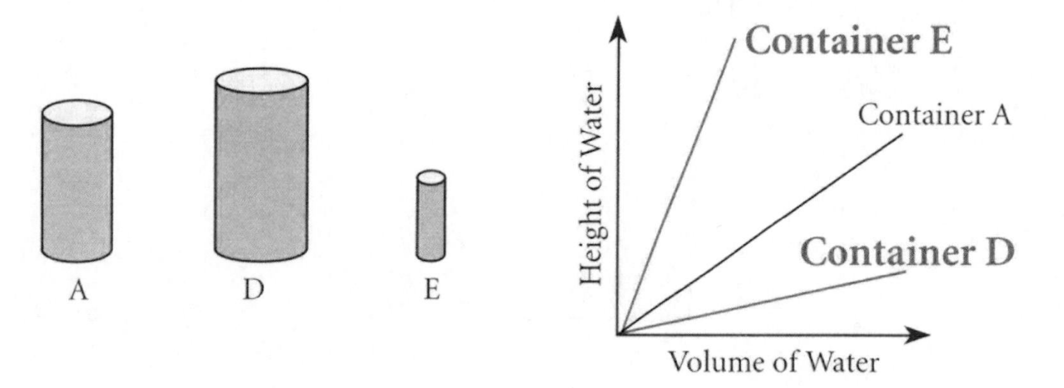

4. Study the graph below. Then draw a line to represent the change in water height as the volume of water increases in container G.

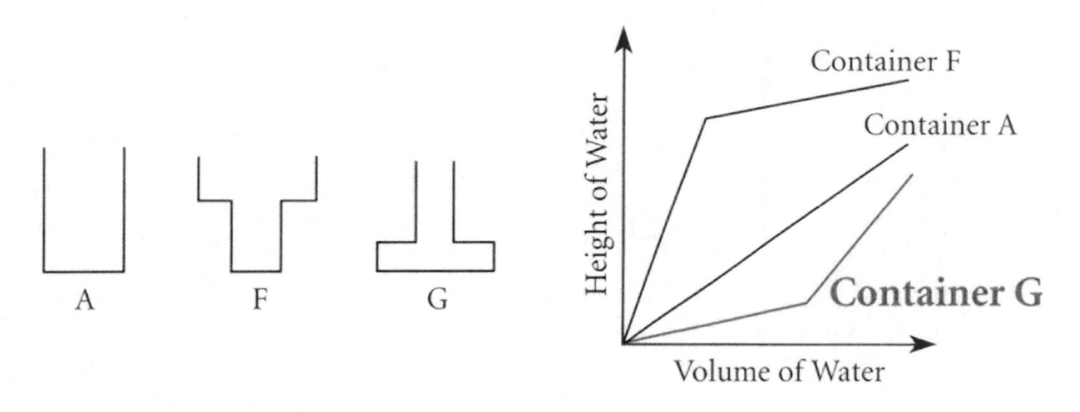

5. Study the following graph carefully. Then make drawings to show what containers H and J might look like.

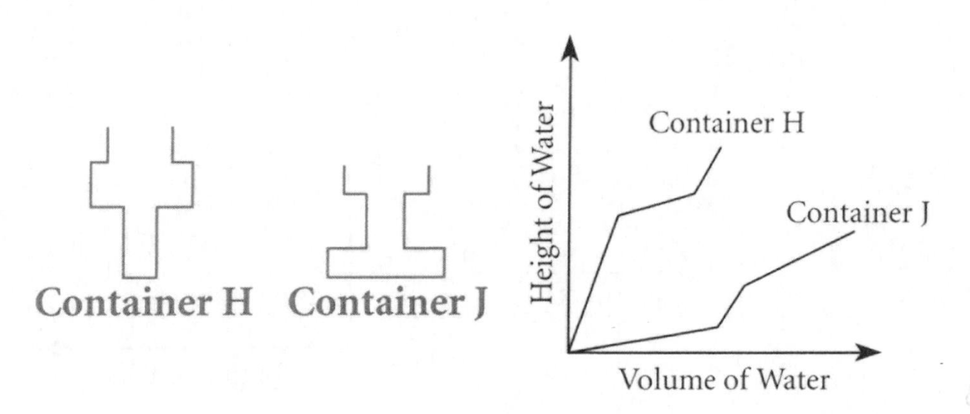

Water is steadily poured into each of these containers. Draw a line on each graph to show how the height of the water changes as the volume increases.

a.

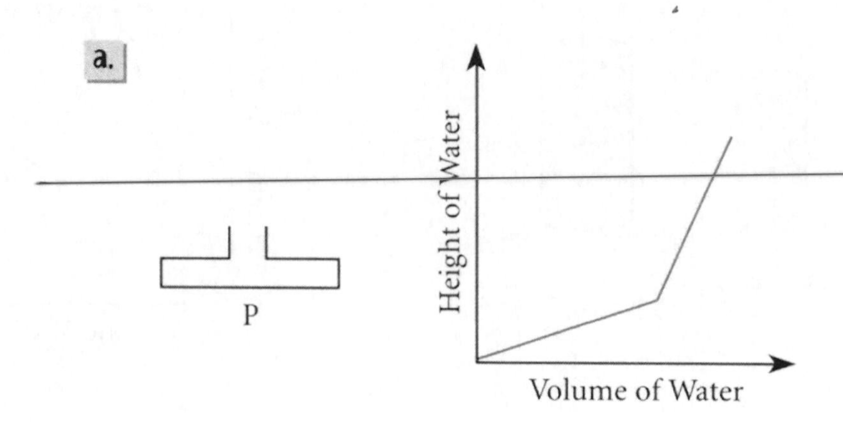

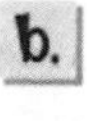

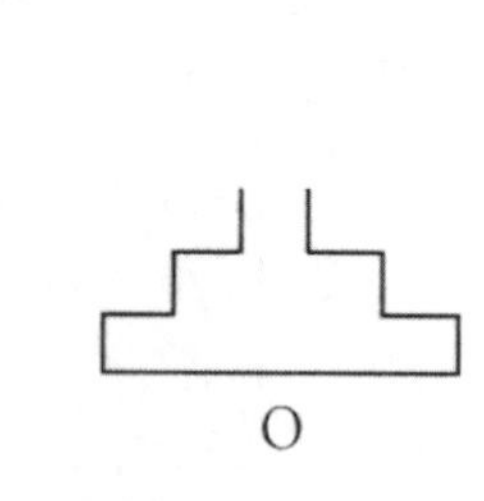

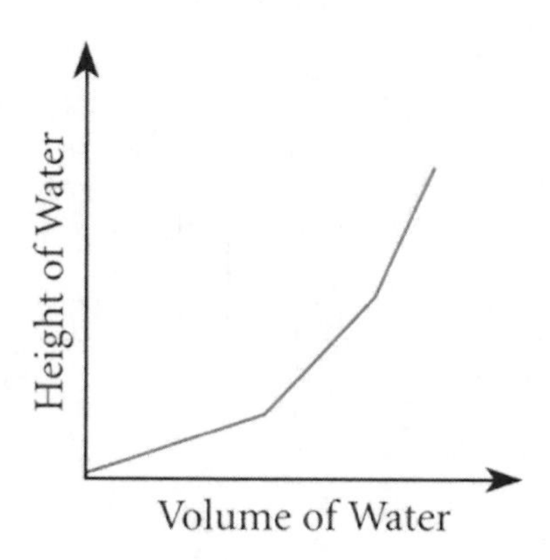

c.

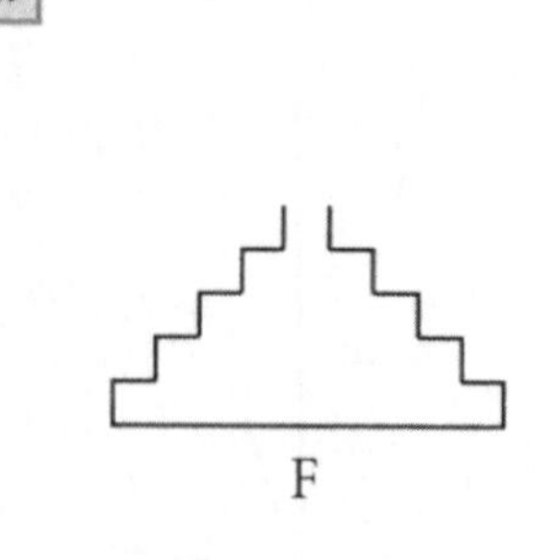

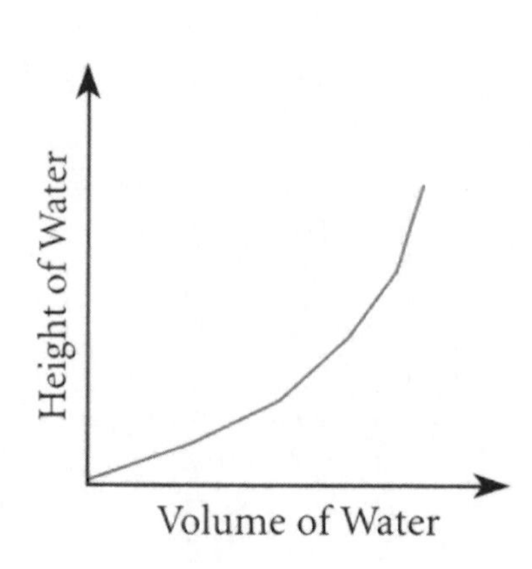

7 Draw a line on each graph to show how the height of water changes as water is steadily poured into each container.

a.

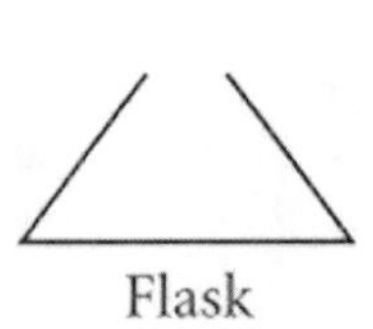

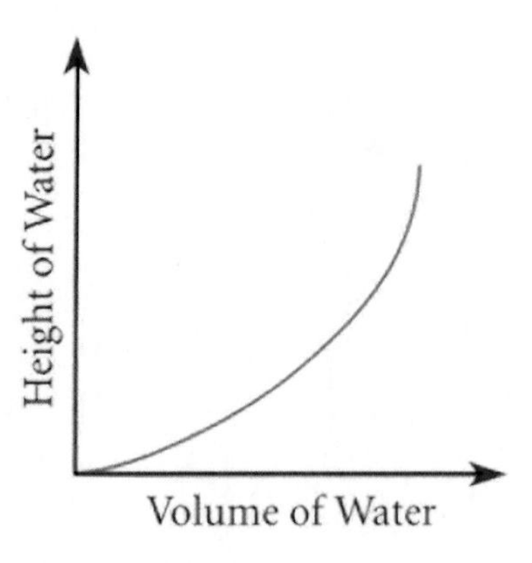

b.

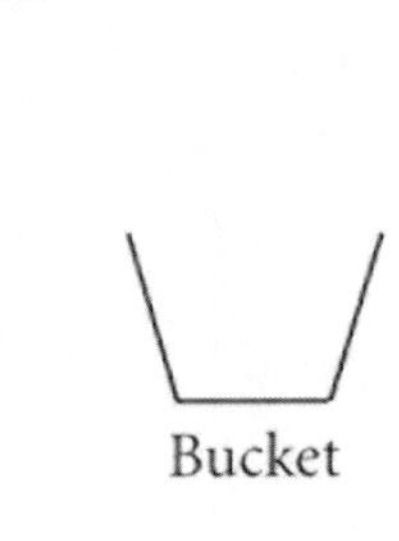

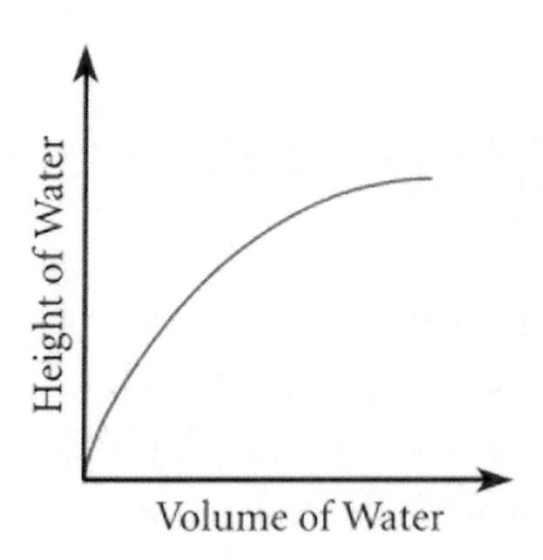

Use after completing page 10.

Write a description comparing the pay rates shown on each graph.

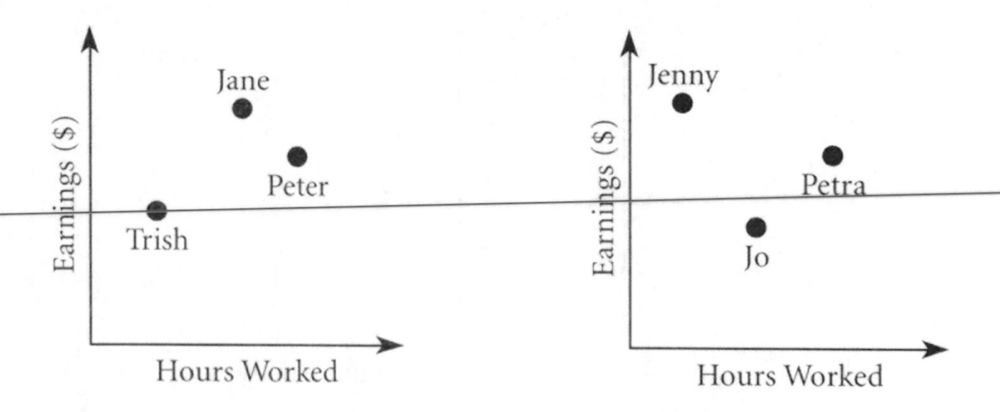

Trish is paid at a higher hourly rate than Jane. Jane is paid at a higher hourly rate than Peter.

Jenny's hourly pay rate is the highest. The hourly pay rates for Jo and Petra are the same.

Use after completing page 19.

This graph shows the height of water in a tank over a period of one month. The water is collected from the rain falling on a roof. Write sentences explaining how the segments of the graph relate to the height of water in the tank and the rate of change.

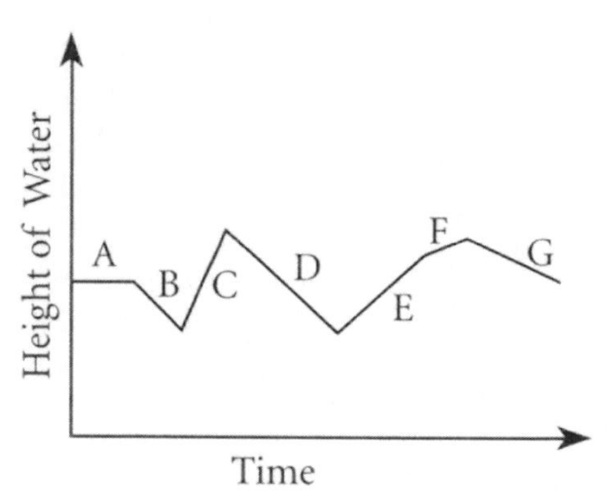

Segment	Explanation
A	The height of water is constant.
B	The height of water is decreasing steadily.
C	The height of water is increasing steadily at a fast rate. (This suggests that it is raining steadily over this period.)
D	The height of water is decreasing steadily at the same rate as it did over the period for B.
E	The height of water is increasing steadily at a rate that is slower than for the period for C. (The line for E is less steep than for C.)
F	The height of water is increasing steadily at a rate that is slower than for any other period of increase.
G	The height of water is decreasing steadily at a rate that is slower than for any other period of decrease.

Practice PLUS 3

Use after completing page 26.

Complete the following conversion diagram for canned vegetables exported to Europe. Then use the diagram to help you complete the table.

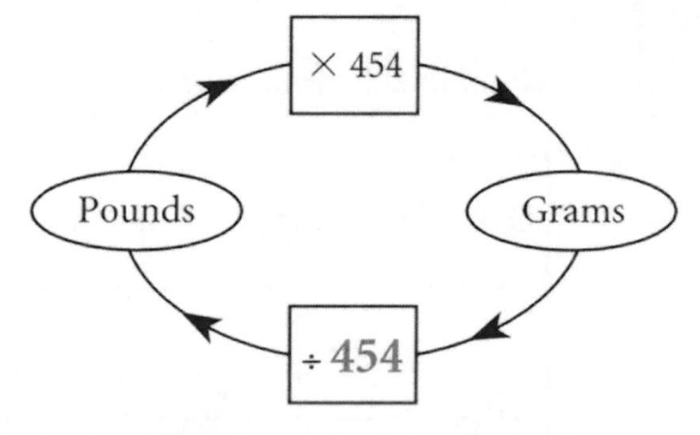

Product	Weight (lb & oz)	Weight (lb)	Weight (g)
Beans	1 lb 4 oz	1.25 lb	568 g
Corn	1 lb 5 oz	1.31 lb	596 g
Tomato Paste	6.4 oz	0.4 lb	182 g
Carrots	9.0 oz	0.56 lb	256 g

Practice PLUS 4

Use after completing page 32.

Use the currency exchange rates to complete the conversion diagram. Then use the diagram to complete the table.

TODAY'S RATES

US$1 = £0.59 (British Pounds Sterling)
US$1 = FF6.41 (French Francs)

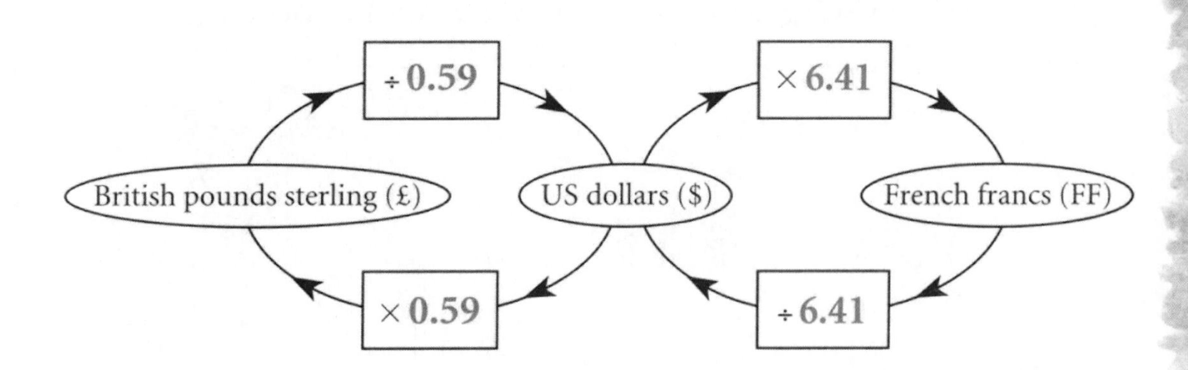

British Pounds (£)	US Dollars ($)	French Francs (FF)
62.54	106	679.46
138.59	234.89	1,505.64
75.23	127.51	817.33
82.61	140.02	897.53

Practice PLUS 5

Use after completing Item 9 page 36.

Complete the following table.

Original Price ($)	Sales Tax	Multiplying Rate	Original Price Plus Tax ($)
85	5%	1.05	89.25
65	3%	1.03	66.95
86	3.5%	1.035	89.01
50	4%	1.04	52
80	2.5%	1.025	82

Use after completing page 41.

Fruit juice is poured into a cylindrical container and also a cone at a steady rate.

Complete each graph to show the height of juice in each container as it is filled.

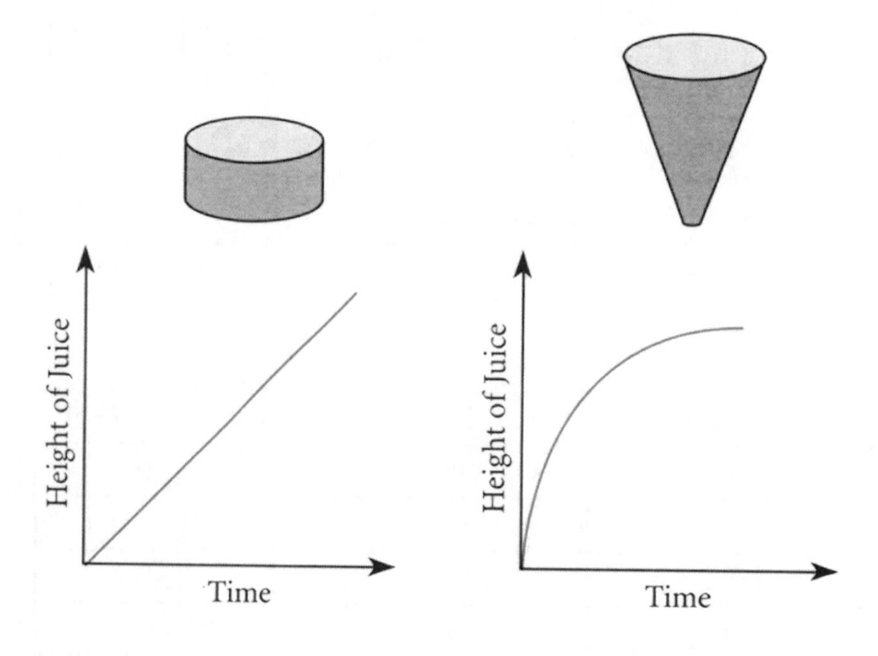

Cumulative Practice

1 Complete each set of number boxes. Numbers are calculated by adding the numbers in the two previous boxes.

2	-1	1	0	1	1	2

-8	3	-5	-2	-7	-9	-16

$1\frac{3}{4}$	$\frac{5}{6}$	$2\frac{7}{12}$	$3\frac{5}{12}$	6	$9\frac{5}{12}$	$15\frac{5}{12}$

-1	$1\frac{1}{6}$	$\frac{1}{6}$	$1\frac{1}{3}$	$1\frac{1}{2}$	$2\frac{5}{6}$	$4\frac{1}{3}$

$2\frac{5}{8}$	-3	$-\frac{3}{8}$	$-3\frac{3}{8}$	$-3\frac{3}{4}$	$-7\frac{1}{8}$	$-10\frac{7}{8}$

-18	$11\frac{1}{2}$	$-6\frac{1}{2}$	5	$-1\frac{1}{2}$	$3\frac{1}{2}$	2

2 Complete each set of number boxes. Numbers are calculated by subtracting the numbers in the two previous boxes.

2	-3	5	-8	13	-21	34

-4	-6	2	-8	10	-18	28

3.2	-3.8	7	-10.8	17.8	-28.6	46.4

6.4	4.8	1.6	3.2	-1.6	4.8	-6.4

$\frac{3}{4}$	$\frac{5}{6}$	$\frac{-1}{12}$	$\frac{11}{12}$	-1	$1\frac{11}{12}$	$-2\frac{11}{12}$

$-2\frac{2}{3}$	-1	$-1\frac{2}{3}$	$\frac{2}{3}$	$-2\frac{1}{3}$	3	$-5\frac{1}{3}$

3 Complete each set of number boxes. Numbers are calculated by doubling the number in the previous box and then subtracting 3.

20	37	71	139	275	547	1,091

$2\frac{1}{2}$	2	1	-1	-5	-13	-29

2	1	-1	-5	-13	-29	-61

$2\frac{29}{32}$	$2\frac{13}{16}$	$2\frac{5}{8}$	$2\frac{1}{4}$	$1\frac{1}{2}$	0	-3

$2\frac{25}{32}$	$2\frac{9}{16}$	$2\frac{1}{8}$	$1\frac{1}{4}$	$\frac{-1}{2}$	-4	-11

$3\frac{1}{8}$	$3\frac{1}{4}$	$3\frac{1}{2}$	4	5	7	11

Find the GCF or LCM.

Numbers	GCF
18, 24	6
72, 96	24
48, 128	16

Numbers	LCM
18, 24	72
72, 96	288
48, 128	384

Calculate.

$\frac{3}{4}$ of $2\frac{1}{2} = 1\frac{7}{8}$

$1\frac{1}{4} \div 3\frac{3}{4} = \frac{1}{3}$

$\frac{1}{4} \div \frac{5}{8} \times \frac{5}{6} = \frac{1}{3}$

${}^-5 - {}^-2(6 - 8) = {}^-9$

$3(7 + {}^-8) - {}^-10 = 7$

${}^-80 \div {}^-2 \times (8 - {}^-12) = 800$

6

$\left(1\frac{1}{2}\right)^2 = 2\frac{1}{4}$

$1^{342} = 1$

$({}^-3)^4 = 81$

$\left(\frac{{}^-2}{3}\right)^3 = \frac{{}^-8}{27}$

$\sqrt{\frac{25}{36}} = \frac{5}{6}$

$\sqrt{\frac{1}{81}} = \frac{1}{9}$

$\sqrt[3]{{}^-343} = {}^-7$

$\sqrt[3]{\frac{1}{27}} = \frac{1}{3}$